KOREA REBORN

A GRATEFUL NATION

HONORS WAR VETERANS FOR 60 YEARS OF GROWTH

Visit us at: www.RememberMyService.com

Library of Congress Cataloging-in-Publication Data

Korea Reborn: A Grateful Nation Honors War Veterans for 60 Years of Growth
Remember My Service Productions
p. cm.

ISBN: 978-0-9863285-0-3

This commemorative book is published and produced with cooperation from the Republic of Korea and the United States of America.

We honor and remember all those who bravely served during the Korean War.

This limited edition commemorative book
was published and produced by
Remember My Service Productions,
a division of StoryRock, Inc.,
in cooperation with Seoul Selection.

RememberMyService.com

In Dedication to All Korean War Veterans

I am greatly pleased to see the publication of this commemorative photo book as we mark the 60th Anniversary of the Korean War Military Armistice Agreement.

This book portrays the brutal scars of war and captures the Republic of Korea arising from such hardship to achieve its remarkable development. Each photo in this book bears witness to the greatness of freedom and democracy and embodies the hearts and souls of the Korean War veterans who fought for freedom and peace.

Sixty years ago, those veterans risked their lives to safeguard freedom in the Republic of Korea. In subsequent years, Korea emerged from the ranks of the poorest countries in the world and has made the unprecedented achievement of both economic development and democratization.

The Republic of Korea, which was once dependent on international aid, is now well-positioned to contribute to others around the world. The blood, sweat, and tears shed by veterans of the Korean War sowed the seeds of today's freedom, peace, prosperity and plenty.

This precious book was published with the combined support of the Korean government, people, and business community and expresses our appreciation to the countries that came to our aid during the War. I hope that this book is passed on to future generations as a book of history that honors the noble sacrifice of the veterans and reminds them of the value of freedom and democracy.

I sincerely thank once again the veterans, their families, and all of our friends from around the world for helping the Republic of Korea become what it is today.

July 2013

Park Geun-hye
President of the Republic of Korea

Department of Defense 60th Anniversary of the Korean War Commemoration Committee

We Remember the Forgotten Victory!

Honoring the Men and Women who Served During the Korean War

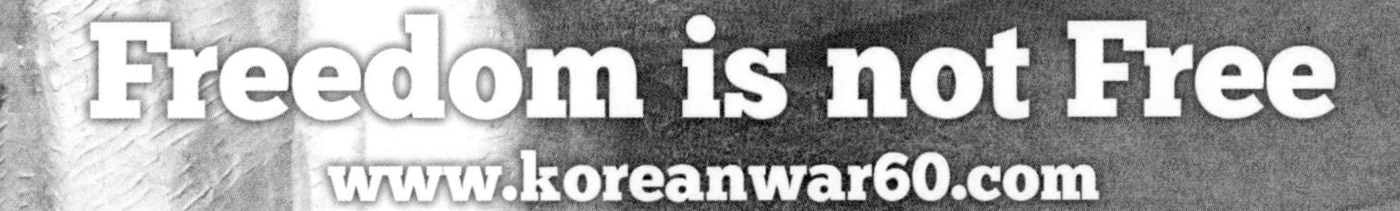

Images courtesy of NARA (left) and Seoul Selection (right)

REBORN
GT
PRUGIO CITY

KOREA REBORN

A GRATEFUL NATION

HONORS WAR VETERANS FOR 60 YEARS OF GROWTH

CONTENTS

PROLOGUE

THE "FORGOTTEN VICTORY"

With her brother on her back, a war-weary Korean girl tiredly trudges by a stalled M-26 tank, at Haengju, Korea. (June 9, 1951, photo by Maj. R. V. Spencer, USAF)

The "Forgotten Victory"

In many ways, the Korean War ended in 1953 the way it had begun three years earlier: as a country divided in half by an imaginary line. Arbitrarily identified as the boundary between North Korea and South Korea at the end of World War II, the 38th Parallel had marked the spot where freedom and communism collide, where a people divided by political and ideological differences meet, and where on June 25, 1950, the Korean War erupted.

For U.N.-sanctioned Allied Forces backing the South (Republic of Korea), the war was over the rights of a people to elect their own leaders, to make their own laws, to choose their own destiny; in short, a war to reject communism and accept freedom. As U.S. servicemen and women descended on a war-ravaged country to support that battle for freedom, they found a nation—and a humble, hardworking people—worth fighting for.

But despite the fact that the war never officially ended, the United Nations Command forces won many victories in the Republic of Korea.

In the past 60 years, the country has changed dramatically. During the war, the GNI per capita was a mere US $67. In 2007, it was nearly US $20,000, with the fastest-growing economy in the world during the 1980s and 1990s. In addition, trade in the country has grown more than 3,000 times since the war, and the Republic of Korea is a rising star among the G-20, a group of the world's leading industrialized nations.

No one can deny that the war certainly laid the foundation for much of this change. In *Korea Magazine* (June 2010), Park Myung-rim, a professor at Yonsei University and an authority on the Korean War, observes: "Through the war, Korea grew up to be much stronger in the long term."

In the same issue of *Korea Magazine*, cultural critic Cho Woo-seok agrees, noting that the Western influence that the U.S. soldiers brought to the country played a significant role: "I'd say one of the factors that helped Korea accomplish modernization relatively fast was the war."

The glimpse South Koreans caught of modernization from their Allied soldier supporters propelled this people beyond their former "hermit kingdom" mentality into a quest for democracy, growth, and success—a quest that would lead them to develop a Korea reborn. ■

TOP: **A refugee family from Masan, living in a refugee camp at Jangseungpo, Korea.** (Photo by the USIA)
BOTTOM LEFT: **A Company Commander gives the communication radio operator the word to move out as this company moved forward during the Korean fighting.** (Photo by the NARA)
BOTTOM RIGHT: **Refugees streaming across the frozen Han River as they flee southward before the advancing tide of Chinese and North Korean Communists. Shattered bridges are shown in the background.** (January 1951, photo by the USIA)

TOP: Long trek southward: Seemingly endless file of Korean refugees slogs through snow outside of Gangneung, blocking withdrawal of ROK 1 Corps. (Photo by Cpl. Walter Calmus, Army)
BOTTOM: Buddies aid wounded man of 24th Infantry Regiment, after a battle 10 miles south of Cheorwon, Korea. (April 22, 1951, photo by Cpl. Tom Nebbia, Army)

ABOVE: A Korean girl places a wreath of flowers on the grave of an American soldier, while Pfc. Chester Painter and Cpl. Harry May present arms, at the U.S. Military Cemetery at Danggok, now officially known as the United Nations Memorial Cemetery Korea. (April 9, 1951, photo by Cpl. Alex Klein, Army)

A Country—and People—Worth Fighting For

Located on a strategic peninsula, Korea is bordered by two powerful countries: China on the northwest and Russia on the northeast. Japan is just east, separated only by the Korea Strait and the East Sea.

Because of its geographic location, Korea has conducted brisk cultural exchanges with its larger neighbors. It has also frequently been the target of aggression. For much of its history, however, the Korean people were successful at fighting off invaders who were interested in controlling the small but valuable piece of land.

In the 1900s, however, that changed. After winning wars against both China and Russia, Japan forcibly annexed Korea and instituted colonial rule in 1910. Japan banned the Korean language, established a cultural assimilation policy, conscripted millions of Koreans into a labor movement, and forced tens of thousands of Koreans into the Japanese military.

This invasion infuriated—and unified—the proud Korean people, who spent decades fighting against Japan. On March 1, 1919, a peaceful demonstration demanding independence spread nationwide, and even beyond the country's boundaries to Manchuria and Siberia. The Japanese authorities stopped the demonstration by massacring thousands, a move that actually strengthened the bonds of national identity and patriotism among Koreans.

Although it failed, the March 1 Independence Movement ultimately led to the establishment of a provisional government in Shanghai and an organized armed struggle against the Japanese in Manchuria.

The Korean people's dogged determination to persevere and fight for what was right continued after the Japanese relinquished control of the country at the end of World War II and became an essential aspect of the Republic of Korea's astounding ability to rebuild after the devastating effects of the Korean War. Strong traditions of hard work, pride, and refusal to give up served South Koreans well as they moved past the war and forward into the next century, clearly showing why they were a country—and people—worth fighting for. ■

A Republic of Korea child sits alone in the street, after elements of the 1st Marine Division and South Korean Marines retake the city of Incheon, in an offensive launched against the North Korean forces in that area. (September 16, 1950, photo by Pfc. Ronald L. Hancock, Army)

TOP: North Korean refugees use anything that will float to evacuate Hungnam. Here they jam the decks of a ROK LST and many fishing boats. (Photo by the NARA)
BOTTOM LEFT: During South Korean evacuation of Suwon Airfield, a weapons carrier hauls a 37-mm anti-tank gun for repairs. (Photo by the USIA)
BOTTOM RIGHT: Refugees crowd a railway depot at Incheon, in hopes that they may be next to get aboard for a trip further south and gain safety from the Communist forces. (January 3, 1951, photo by C. K. Rose, Navy)

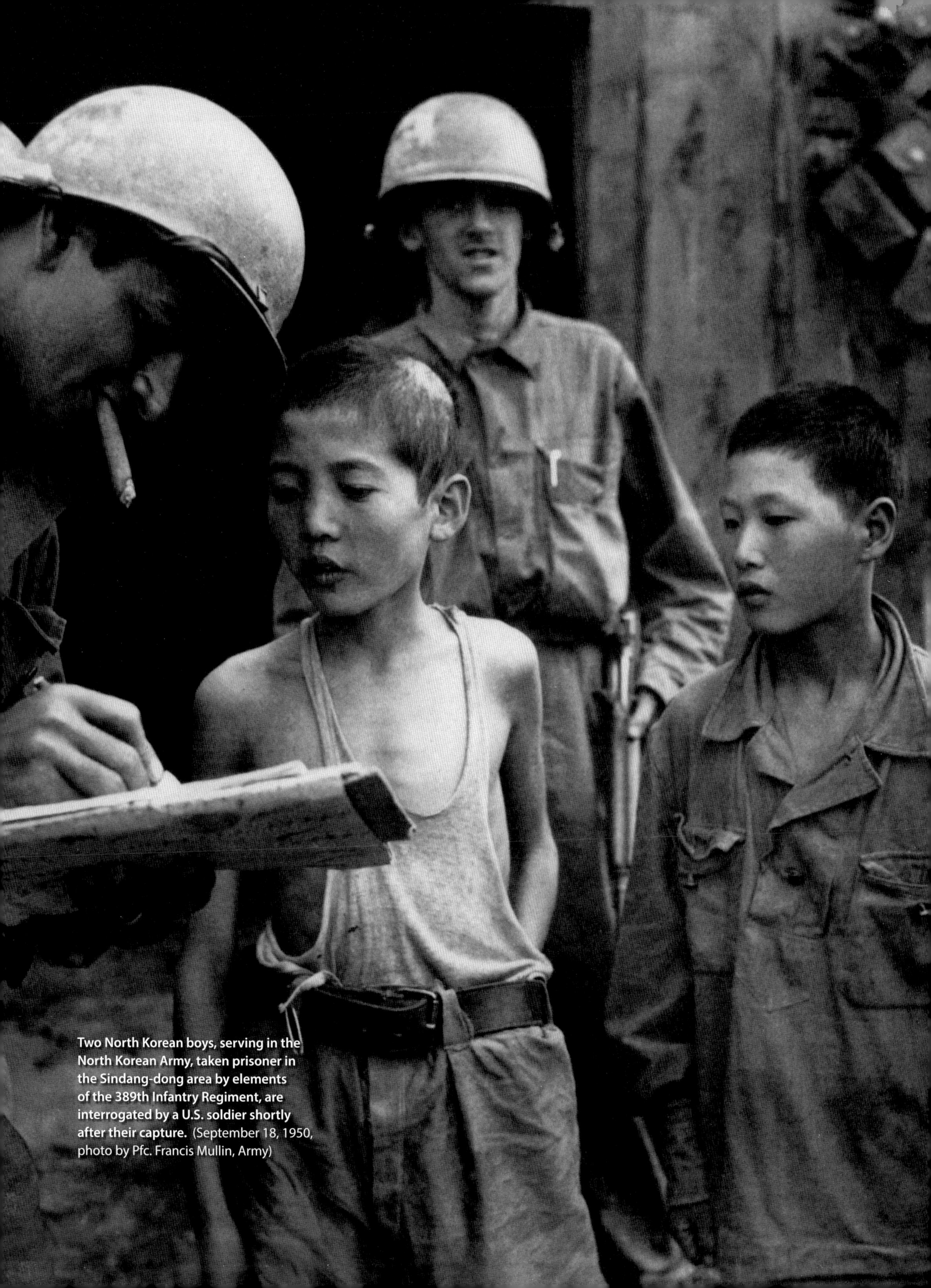

Two North Korean boys, serving in the North Korean Army, taken prisoner in the Sindang-dong area by elements of the 389th Infantry Regiment, are interrogated by a U.S. soldier shortly after their capture. (September 18, 1950, photo by Pfc. Francis Mullin, Army)

A grief-stricken American infantryman whose buddy has been killed in action is comforted by another soldier. In the background, a Corpsman methodically fills out casualty tags, Hakdong-ni area, Korea. (August 28, 1950, photo by Sgt. 1st Class Al Chang, Army)

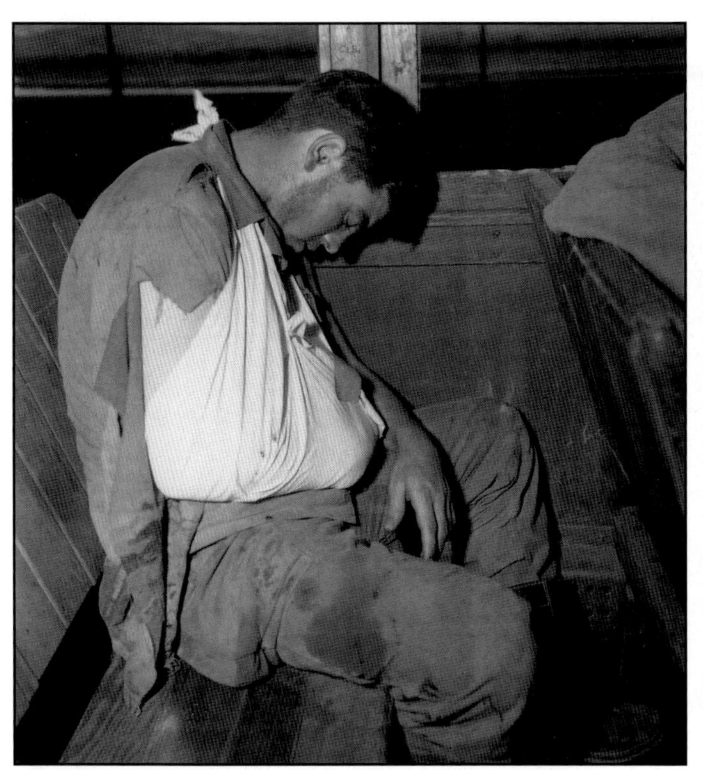

Pfc. Orvin L. Morris, 27th Regiment, takes a much-deserved rest during his evacuation to Busan, Korea, on a hospital train. He was wounded by enemy mortar fire on front lines.
(July 29, 1950, photo by Sgt. Dunlap, Army)

Troops are climbing down a cargo net to waiting LCVPs as they land. (January 9, 1953, photo by the Army)

The Best Ending

This is a portion of a winning essay about the Korean War, written by Karen L. Healy, daughter of former Army Master Sgt. Jason Walker Bowling.

Dad used to talk to us about the Army . . . his Army. Over a crowded dinner table, during long vacation drives to Tennessee or Kentucky, sometimes even when he just wanted to make a point about how easy life was for my generation, he would talk about his Army. More specifically, he would talk about the time he spent in Korea. . . .

Two years ago, my younger sister and I took Dad and his wife to Washington, D.C. I had to be there on business; Dad wanted to see the Korean Memorial. . . . At the Memorial, we hung back and let him absorb the entire thing at his own pace. . . . Dad walked gingerly over the long marble slabs — the artist's rendering of the paths in a rice paddy. Dad looked into the face of nearly every one of the larger-than-life soldiers. I think he saw people he knew in each of their faces. He walked the entire line with the soldiers, I think on patrol in some far-off field in his own mind. . .

The following day, I had arranged for lunch at the National Press Club, courtesy of a colleague who had joined Kia Motors the year before. . . [and] we were joined by some of the company's Korean executives.

Once the introductions and pleasantries were over, a gentle man with a soft voice stood and addressed my father. He talked about life in Korea when he was young. He talked about the American soldiers and the conflict to create a free country. He talked about his ability today to work for a great company, now doing business in an even greater country. He told my dad that none of that would have been possible, and that his life would have been entirely different, if the American soldiers had not given his country freedom. And he shook my Dad's hand and thanked him for his service.

I never understood until that moment the bond between my Dad and Korea. He wasn't there very long, but he left part of himself there that he'll never get back. And he brought back a lot of Korea in his heart. He still tells stories on occasion, but I'm sure none of us will ever hear the realities of what he lived through. It's probably still too painful. But the joy on his face as he shook the hand of the soft-spoken Korean man was the best ending to any of his stories. ■

TOP: A Korean family mourns their murdered father, victim of the wholesale murder at Jeonju by North Koreans. (September 27, 1950, photo by Master Sgt. E. T. Tarr, Army)
BOTTOM: Lt. Col. John Hopkins, of the First Battalion, Fifth Marine Regiment, leads in singing the "Star-Spangled Banner" during memorial services held in the field during the Korean campaign. (June 21, 1951, photo by Cpl. Valle, Marine Corps)

U.S. Marines wounded at Mt. Garisan are evacuated via helicopter and flown to a hospital in the rear area for treatment. Navy Corpsmen prepare three wounded Marines for evacuation. (May 23, 1951, photo by N. H. McMasters, Navy)

ABOVE: Pfc. Dwight Exe, 5th Cav. Regiment, catches up on his letters to the folks at home during a break in action against the Chinese forces along the fighting front in Korea. (November 15, 1951, photo by Cpl. James L. Chancellor, Army)

American and ROK soldiers team up in a position to battle somewhere in Korea. (July 20, 1950, photo by the NARA)

ABOVE: With nearly 3,000 pin-ups serving as wallpaper for their Quonset hut, these Marines of the Devil Cats Squadron are still looking for more. (October 28, 1952, photo by Sgt. Curt Giese, Marine Corps)

RIGHT: Marilyn Monroe sings several songs for an estimated 13,000 men of the First Marine Division. Miss Monroe stopped at the First Marine Regiment on her tour of the military units in Korea.
(February 16, 1954, photo by Cpl. Kreplin, Marine Corps)

ABOVE: Retired General Paik with Korean War Veterans

"…I shall be at the front. If I turn back, shoot me."

At the outset of the Korean War, a then-29-year-old Colonel Paik distinguished himself as the commander of the ROK 1st Infantry Division. During the battle at Dabudong (a.k.a. the "Bowling Alley"), Paik's 1st Division bravely counterattacked a ridgeline previously lost to the attacking North Korean Army. Paik famously inspired his troops by telling them, "We are going to turn around and kick the enemy off our ridge, and I shall be at the front. If I turn back, shoot me."

Being at the forefront of leadership has always been a trademark of Paik Sun Yup. In November 1951, the ROK 1 Corps (later named Task Force Paik) mounted a campaign against guerilla activity in the Mt. Jirisan region of southwestern Korea. Dubbed Operation "Rat Killer," the troops under Paik's command (now a General) captured or killed an estimated 25,000 guerillas by March 1952.

In July of the same year, General Paik was appointed ROK Army Chief of Staff, the highest position in the ROK Army. At only 32-years old, he commanded ten Army divisions, which would grow to 20 divisions by 1953.

General Paik Sun Yup became Korea's first officer to attain four-star rank, and would later command the First Field Army, serve a second appointment as Army Chief of Staff, and finally serve the remainder of his career as Chairman for the ROK Joint Chiefs of Staff. He retired from military service in 1960 and began a second career as a diplomat, serving as ambassador to China, France, and Canada. Following his diplomatic service in 1969, he served as Minister of Transportation until 1971. Thereafter, he served as president of two national policy companies.

Today, General Paik is a distinguished author and respected voice for Veterans' affairs. With a lifetime devoted to public service and his family, General Paik Sun Yup remains an iconic figure in South Korean history.

TOP: Wounded American soldiers are given medical treatment at a First Aid station, somewhere in Korea.
(July 25, 1950, photo by Pfc. Tom Nebbia, Army)
MIDDLE: Marine Corps tanks—ready for the front lines—are swung aboard a barge at the Naval Supply Center by crane, for transshipment to UNC forces in the Pacific Far Eastern Command.
(1950, photo by the USIA)
BOTTOM: A U.N. LST slips into the harbor at Incheon following its amphibious landing by U.S. Marines. (December 13, 1950, photo by the Navy)

Map of the Korean War

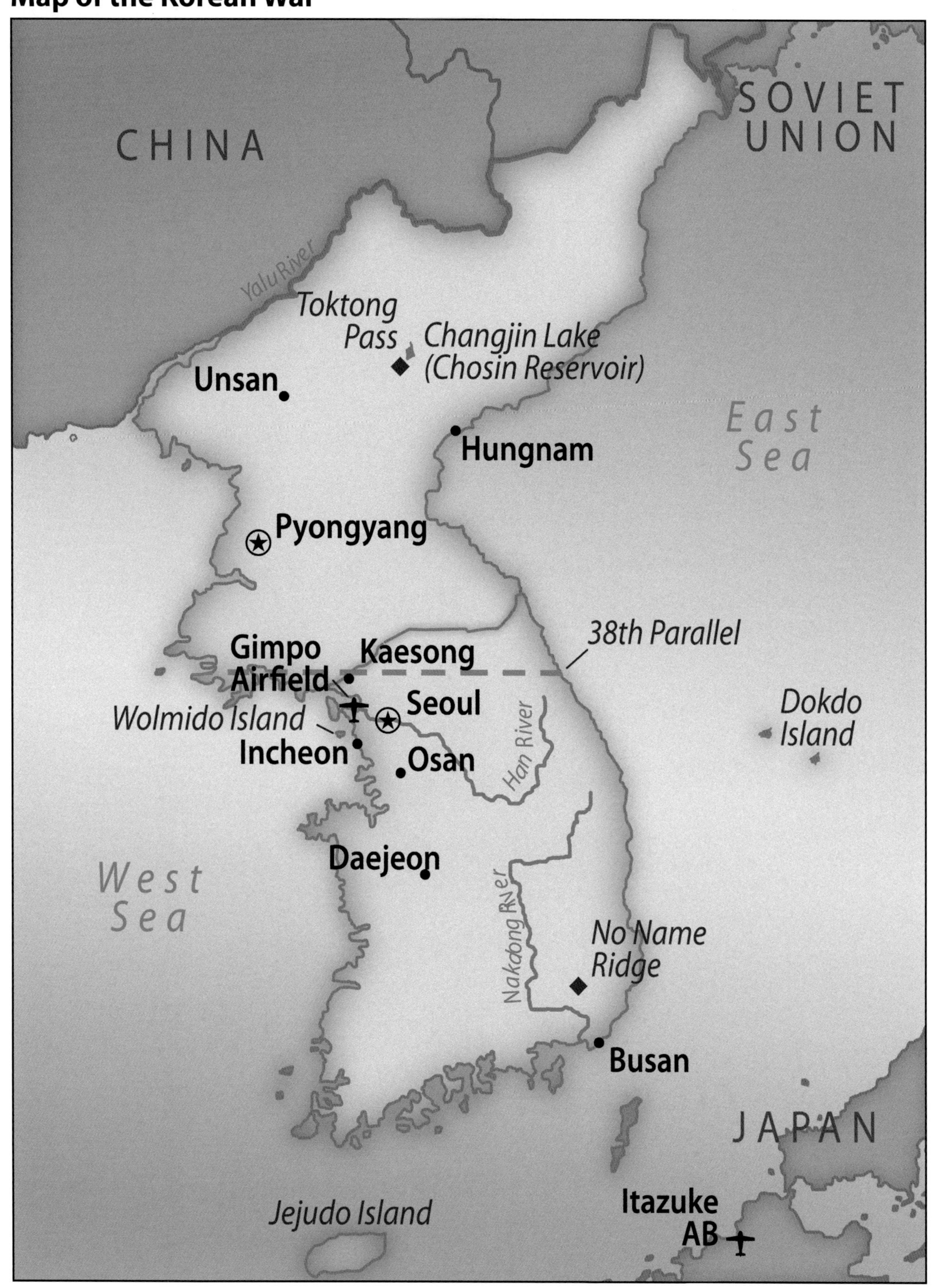

SECTION I

THE ATTACK

1

Korea becomes an independent nation. President Syngman Rhee speaks at the ceremonies of the inauguration of the newly formed government of the Republic of Korea. (Seoul, Korea, August 15, 1948, photo by the NARA)

CHAPTER 1

Setting the Stage

When World War II ended in 1945, few realized that another war was beginning. It would take five years for the conflict in Korea to gain international attention—and participation—but it started when the Empire of Japan, which had occupied Korea for more than three decades, was defeated in World War II by the Allies. In order to accept the surrender of Japanese forces in Korea, the Americans occupied Korea south of the 38th Parallel, and the Soviet Union occupied the peninsula north of the Parallel.

In theory, the two countries agreed to temporarily occupy the country as "trustees," establishing a Korean provisional government that would ultimately lead to a free and independent country.

Wary of the "domino effect" (the concern that if one state in a region were controlled by the Communists, surrounding countries would soon follow), the United States began to watch Korea closely. President Harry S. Truman persuaded the United Nations to assume responsibility for Korea. The United Nations called for general and free elections to be held in both halves of Korea. The Soviet Union refused to cooperate, and the die was cast: a Communist state with Soviet support was established in North Korea, while a democratic movement gained favor in South Korea.

Kim Il Sung, a major who had led a Korean contingent in the Soviet army, returned to North Korea from Manchuria to lead the new Communist government. With the backing of both the Soviet Union and China, he planned to unify the two Koreas by military force. Meanwhile, under U.N. mandate, the democratic Republic of Korea was being formed in the south, with Syngman Rhee elected as its first president. This move toward freedom hadn't come easily, however; some quarters of Southern society opposed the new state, resulting in upheavals throughout the country.

The end result? North Korea's military forces and supplies grew substantially, while in South Korea, both human and equipment resources were dramatically depleted in the civil unrest. Although there were attempts to unify the two countries, tension grew and cross-border raids and skirmishes at the 38th Parallel became commonplace. When North Korea launched an attack on June 25, 1950, the Republic of Korea forces were under-prepared, under-supplied, and outnumbered. ■

TOP: Brig. Gen. Courtney Whitney; Gen. Douglas MacArthur, Commander in Chief of U.N. Forces; and Maj. Gen. Edward M. Almond observe the shelling of Incheon from USS *Mt. McKinley*. (September 15, 1950, photo by Nutter, Army)
BOTTOM: Jacob A. Malik, Soviet representative on the U.N. Security Council, raises his hand to cast the only dissenting vote to the resolution calling on the Chinese forces to withdraw troops from Korea. (Lake Success, NY, December 1950, photo by the USIA)

The Beginning of the Cold War

The term "cold war" was first used by English author and journalist George Orwell in an essay titled, "You and the Atomic Bomb," published October 19, 1945, in the British newspaper *Tribune*.

Orwell wrote about the impact of the threat of nuclear war, predicting a nuclear stalemate between "two or three monstrous super-states, each possessed of a weapon by which millions of people can be wiped out in a few seconds... Few people have yet considered its ideological implications—that is, the kind of world-view, the kind of beliefs, and the social structure that would probably prevail in a state which was at once unconquerable and in a permanent state of 'cold war' with its neighbors."

Orwell used the term a second time when writing an article for *The Observer*, which was published March 10, 1946. In that article, Orwell observed that "after the Moscow conference last December, Russia began to make a 'cold war' on Britain and the British Empire."

In 1947, American financier and presidential advisor Bernard Baruch delivered a speech where the term was used to describe the post-World War II tensions between the United States, its Western European allies, and the Soviet Union: "Let us not be deceived: we are today in the midst of a cold war."

That same year, the term became more widely used and recognized when Walter Lippman, a newspaper reporter, wrote a book called *The Cold War*.

It is generally accepted that the Cold War began near the end of World War II, with an uneasy truce declared between the United States and the Soviet Union. The United States and its Western European allies were committed to a system where individual countries were led by democratic governments, and differences were resolved by international organizations. With a history of invasion and incredibly high casualties, the Soviet Union wanted to increase security by controlling the governments of the countries that bordered it. The United States and its Western European allies watched warily as the Soviet Union began to establish Communist governments in several Eastern European countries that had been liberated by the Red Army.

The Cold War reached its peak in the years preceding and throughout the Korean War. During this time, the Soviets unsuccessfully blockaded the Western-held sectors of West Berlin (1948–49); the United States joined with its European allies to create the North Atlantic Treaty Organization (NATO), a unified military command aimed to resist the Soviet presence in Europe (1949); the Soviets exploded their first atomic warhead, ending the American monopoly on the atomic bomb (1949); and a Communist government came to power in China (1949). Finally, Soviet-backed North Korean armies invaded South Korea, launching the Korean War.

ABOVE: President Harry S. Truman is shown at his desk at the White House signing a proclamation declaring a national emergency in Korea. (December 16, 1950, photo by the USIA)

The sustained state of political and military tension between the United States and Soviet Union continued for several decades. The Cold War was waged largely on propaganda, economic, and political fields, but military clashes—largely indirect, such as in Vietnam and Afghanistan—took place, too. In 1991, the Soviet Union collapsed, as its various satellite states rose in mostly peaceful revolts, leaving the United States as the sole superpower. ■

ABOVE: United Nation Forces recross the 38th Parallel ahead of pursuing Chinese forces. (1950, photo by USIA)

Choosing the 38th Parallel

As World War II drew to an end, leaders of the winning countries (United States, United Kingdom, and the Soviet Union) gathered together at the Potsdam Conference to discuss the details of a peace treaty officially ending the war. As part of that treaty, the 38th Parallel (a popular name given to 38° N Latitude) was chosen as a separation between South and North Korea.

The line roughly divides the peninsula in half. According to the agreement, the Soviet Union was to accept the surrender of Japanese forces north of the line, while the United States was to accept the Japanese surrender south of the line.

Initially, the 38th Parallel was intended to be a temporary division of the country, while the United States and the Soviet Union assisted the two Koreas in establishing one country and one government. The onset of the Cold War, however, created entirely new circumstances, which ultimately led to the Korean War. ■

TOP LEFT: Korean refugees prepare to board an LST during the evacuation of Hungnam, while other refugees unload some of their meager belongings from an ox-cart and load them on a fishing boat. (December 19, 1950, photo by the Navy)
TOP RIGHT: Miss Mo Yun Sook, famed Korean poet, telling how she escaped the Communist-led North Koreans when they captured Seoul, by hiding in the mountains until the U.N. Forces liberated the city. (November 8, 1950, photo by Cpl. Robert Dangel, Army)
OPPOSITE PAGE TOP: General Douglas MacArthur inspects troops of the 24th Infantry on his arrival at Gimpo Airfield for a tour of the battlefront. (February 21, 1951, photo by the USIA)
OPPOSITE PAGE BOTTOM: United Nations flag waves over crowd waiting to hear Dr. Syngman Rhee speak to the United Nations Council in Daegu, Korea. (July 30, 1950, photo by Sgt. Girard, Army)

2

Cpl. John W. Simms of Bradbury Heights, Maryland, bids his wife, Ann, and their 9-month-old son, John Jr., goodbye as he leaves for Korea.
(1950, photo by the Washington Post, USIA)

CHAPTER 2

The War Begins

As early as 1949, Kim Il Sung had approached Soviet leader Joseph Stalin, asking for support to invade the South. Unconvinced that North Korean forces were prepared, and concerned about the response of the United States, Stalin refused. During the next year, however, numerous Korean veterans returned from China after serving in the People's Liberation Army, strengthening North Korean forces. By mid-1950, Stalin was on board, and the Soviets provided additional armaments.

Before the sun rose on June 25, 1950, North Korea began a thunderous artillery attack along the 38th Parallel. More than 50,000 soldiers poured across the Imjin River, heading toward Seoul. Another 54,000 soldiers attacked several cities located strategically along the way.

It took only three days for the North Korean troops to reach their destination, wreaking havoc and destruction in their wake. Once in Seoul, however, they were unexpectedly stopped—temporarily. The badly beaten Republic of Korea forces made a desperate stand to protect their capital city, forming a defensive line along the Han River. Knowing they couldn't hold off the enemy for long, the Rhee government sent out a frantic plea for help to the United States.

Initially, President Truman ordered General Douglas MacArthur to oversee operations to provide the Republic of Korea with munitions and evacuate U.S. citizens. Pres. Truman also turned to the United Nations, which had called for the invasion to stop the day it started.

The United Nations Security Council passed a resolution calling on member states to provide military assistance. As a permanent member of the United Nations, the Soviet Union could have vetoed the resolution, but the country had boycotted the Council after the United Nations recognized the Republic of China, which had relocated to Taiwan after the Chinese Civil War. Because the Soviets weren't there, the resolution passed with little resistance.

The United States quickly deployed nearby military forces stationed in Japan, but the lack of preparation was evident. U.S. soldiers fought valiantly alongside their Republic of Korea counterparts, but inadequate weapons and manpower meant the North Koreans continued their relentless advance southward into the Korean peninsula.

A stunned world—including an American public—watched as a fledgling democratic country, along with its U.N. allies, came dangerously close to being destroyed. ■

Sgt. Jim Ecerett, of the Lowell Main Recruiting Station, gives prospect Carville Berehman the R-2 and R-3 enlistment papers. (June 12, 1950, photo by the NARA)

Signing Up to Fight

On a wave of post-patriotic enthusiasm following victory in World War II, a new generation of young men and women were eager to fight in the Korean War. Some were drafted; others signed up. Many had little or no idea what they were getting into—they just knew they wanted to serve.

"Not quite a year after we graduated from high school, a whole group of us got drafted at the same time," recalls Bill Hartsock, a sergeant in the U.S. Air Force. "We were bussed up to Des Moines, and we all took our physicals. They told us that if we wanted to enlist in something other than the Army, we would have 30 days to make the choice. Some of us flipped coins... that's how I ended up in the Air Force."

Willie B. Harris, who retired as a Command Sergeant Major in the Air Force, enlisted in 1949 because it was a tradition. "I had three brothers in World War II," he notes, "and when I was old enough to realize, you know, I had a dream to be an airborne soldier." Harris recalls that his first experience "wasn't very nice. The second day I was detailed for KP, which is kitchen police, and I must have peeled about 2,000 pounds of potatoes."

Robert L. Cornwell was a senior in high school when he and his buddy started talking about serving. "We decided we might like to go to the Navy. Then one Saturday he came in and I was playing pool, and he said, 'Let's go to Burlington and enlist.' I said, 'OK,' and so we went." After boot camp, Cornwell was assigned to an LST, "an amphibious ship that will run up onto the beach, drop a ramp, and unload directly onto the beach." He immediately got seasick. Despite that shaky beginning, Cornwell earned several Korean service medals. "One of my most memorable experiences was when we went above the 38th Parallel, into enemy territory, and spent two weeks being a target and trying to draw fire from the shoreline so we could find out where they were." ■

Log-lifting combines physical exercise with team work. Recruits are kept outdoors most of the time. (September 16, 1950, photo by the NARA)

Boot Camp: An Introduction

Boot camp is a universal military requirement regardless of the branch of service. The intense introduction to military life is an experience no service member forgets.

"When I was in boot camp at Parris Island, they were really adamant about everybody being a good marksman with the M-1 rifle. Of the 10 or 11 weeks, three of those weeks were solid weeks spent out on the rifle range," remembers Rexford Early, who served as a Marine during the war. "The first week just snapping in the different positions and blank firing. The second week we actually got live ammunition and you practiced and shot, and at the end of the third week, everybody had to qualify. We had to rapid fire, we had to slow fire, and the last 10 rounds of qualification was 500 yards, now that's five football fields. If you didn't qualify, they made you march all the way back to camp... about eight or ten miles."

Early continues, "I don't care how cocky you are, how tough you are, how smart you think you are, they whittle your ass down to nothing. They make you think you are the lowest thing that ever walked on the face of the earth... But after a few weeks, they start building you up. By the time you graduate, you think you're the toughest, meanest son of a gun that ever wore a pair of low cuts. And that's the psychology... you're going to do things that you might think are humanly impossible or you might be scared to death, but you're going to do them because you're not going to let your buddies down.

The most memorable moment of the war for Early was graduating from boot camp. "My DI, who had kicked me, hit me, cursed at me, yelled at me, and threatened me, stuck out his hand, pinned the globe and anchor on my collar and said, 'Congratulations, you are now a Marine, and I would be proud to serve with you.' That was a hell of a compliment... that was the proudest moment of my life." ■

TOP: Pfc. Paul Rivers, 23rd RCT, 2nd Infantry Division, looks for enemy snipers in a burning village as U.S. troops launch an offensive against the North Korean forces in the Yeongsan area. (September 16, 1950, photo by the NARA)
BOTTOM: The Hon. S.Y. Lee, Vice President of the Republic of Korea, leads cheers at the close of the U.N. Day ceremony at Seoul. (October 24, 1950, photo by Sgt. Ray Turnbull, Army)

A multitude of U.S. troops on a pier after debarking from their transport ship, somewhere in the Republic of Korea. (August 6, 1950, photo by Sgt. Dunlap, Army)

The UN's First Military Deployment

South Korea and its allies fought under the UN flag. The United States was, by far, the largest non-Korean contributor of troops to the UN side, but 15 other nations also dispatched troops. The following nations contributed combat forces:

> United States, United Kingdom, Australia, Netherlands, Canada, New Zealand, France, Philippines, Turkey, Thailand, Greece, South Africa, Belgium, Luxembourg, Columbia, Ethiopia.

Sweden, India, Denmark, Norway and Italy also contributed medical support units. Non-U.S. forces played leading roles in many of the Korean War's most critical battles. For instance, the dramatic last stand of the British Gloucestershire Regiment during the Battle of Imjin River and the Australian, New Zealand, and Canadian victory at the Battle of Gapyeong in April 1951 gave UN forces the time they needed to retreat to prepared positions north of Seoul, saving the Korean capital from capture by the Communist Chinese. ■

TOP: Supplies and equipment are evacuated from advancing Chinese forces bearing down on Hungnam, Korea. (December 11, 1950, photo by Pfc. Emerich M. Christ, Army)

BOTTOM LEFT: Rain fails to halt the men of the Heavy Mortar Company of the 5th Marine Regiment, 1st U.S. Marine Division, as they fire 4.2 inch mortars at enemy-held positions, during the assault against the Chinese forces. (Photo by the NARA)

BOTTOM RIGHT: U.S. Marines stand along the rail and watch the ocean aboard USS *Clymer*.
(July 1950, photo by Sgt. Frank C. Kerr, Marine Corps)

3

Fresh and eager troops, newly arrived at the vital southern supply port of Busan, prior to moving up to the front lines.
(August 1950, photo by the USIA)

CHAPTER 3

Standing Strong in Busan

It took only a few weeks for the North Korea People's Army to occupy almost all of South Korea. The mood was frantic when U.N. troops dug in their heels at the Busan Perimeter. This desperate stand at Busan gave the U.N. forces invaluable time to gather the men, equipment, and political support necessary to not only stop the southern progress of the North Koreans, but to begin moving northward and recapturing lost ground, including Seoul.

U.N. troops, led by Lieutenant General Walton Walker, had endured weeks of heartbreaking losses. With time, Lt. Gen. Walker was hoping to build up stronger forces so he could mount an offensive against the enemy, but in Busan, time ran out. The U.N. forces knew they had to hold on to Busan, the last open deepwater port in the Republic of Korea. Vital manpower and equipment from the United States and Japan were arriving by ship daily, and critical airfields were also located there.

On July 29, Walker issued a "stand or die" order, stating, "There will be no more retreating, withdrawal, or readjustment of the lines or any other term you choose… I want everybody to understand we are going to hold this line."

Forces came mostly from the U.S. Marine Corps, the U.S. Army, the Republic of Korea, and the British Army. The desperate, heroic battle continued all along a 140-mile defensive line that protected the vital port of Busan. The battle became known as the Battle of the Busan Perimeter.

For six weeks, these stubborn soldiers refused to give in, fighting off repeated attacks. By this time, North Korean troops were spread thin along a line of devastation from the 38th Parallel, through Seoul, and south into the perimeter. Their supplies were running low, and they had suffered terrible casualties. Nevertheless, they continued to attack, launching a carefully planned offensive that included simultaneous attacks in as many as five locations.

However, things were finally coming together for the U.N. forces, which for the first time enjoyed an advantage in troops, equipment, and logistics. The U.N. troops broke out all along the front on September 16, the day after the U.S. X Corps made a surprise landing at the port of Incheon and quickly captured the city just 35 miles west of Seoul. The Battle of the Busan Perimeter was over. Within days, the shattered North Korean forces were retracing their steps, heading north in defeat, with U.N. soldiers in hot pursuit. ■

Men of the 9th Infantry Regiment ride on an M-26 tank to await an enemy attempt to cross the Nakdong River. (September 3, 1950, photo by Cpl. Thomas Marotta, Army)

View of Busan City and harbor with hospital ship at dock. (September 29, 1950, photo by the NARA)

One Soldier—All the Protection We Had

Bernadette Reider had received basic nursing training when she signed up for the Army Nurses Corps. "I already had two brothers in the service, so I decided to go and do what I could," she recalls. She served in the Korean War, working at hospitals near Busan: "I remember arriving in June of 1950, and we were the first hospital there in Busan. We were on a ship that had arrived from Japan, and there was one man, a soldier sitting up on a hill there with a machine gun, and that was it, that was all the protection we had. So then we disembarked and set up our hospital in an old school, and I'll tell you it was full of fleas. Everybody was just covered with bites; we stayed there a couple of weeks, and then we moved to another building which was a lot better.

"My most memorable experience was when another nurse and I took a train up country to pick up the wounded. It took us two days to get there; we had one car hitched onto the engine. My goodness, when we got there, everyone was running around, and all we had time to do was run from our train into the other one, and then we took off. The North Koreans were almost there, and if we hadn't gotten there when we did, we probably wouldn't have gotten out. We had no medical supplies on the train, and there was nothing we could do... We had five cars full of wounded... the fighting was so heavy, and there were so many casualties; they just kept coming and coming." ■

LEFT TOP: Members of the 17th RCT, composed of U.S. troops and ROK troops, exercise at the Busan Docks, Korea, during a short holdover there before moving to the front lines. (September, 1950, photo by the NARA)
LEFT MIDDLE: 30 July 1951 - Capt Clifford McKeon, 2nd Logistical Command, hands out footballs that were sent to the Happy Mountain Orphanage, Busan, Korea. (Photo by the NARA)
LEFT BOTTOM: Marine Sgt. Charles R. Hill takes aim at those who tried to escape the fury of the Marine advance in Korea. (Photo by the NARA)

31 July 1950 - Members of the 89th RCT disembark from USNS *Gen. K.K. Patrick* at Busan, Korea.
(Photo by the NARA)

4

As against "The Shores of Tripoli" in the Marine Hymn, Leathernecks use scaling ladders to storm ashore at Incheon in the amphibious landing on September 15, 1950. The attack was so swift that casualties were surprisingly low. (Photo by Staff Sgt. W.W. Frank, Marine Corps)

CHAPTER 4

The Incheon Landings

Combined with the breakout from the Busan Perimeter, the daring landing at Incheon, a port near Seoul, marked the turning of the tide in the Korean War. With this successful surprise landing, U.N. forces began a strong push eastward, first recapturing Seoul, then continuing to pursue retreating North Korean troops.

General MacArthur had started planning a landing behind enemy lines in early July and set a target date of September 15. Concerns about the landing were valid: military chiefs of staff worried that by dividing troops between Incheon and the Busan Perimeter, defeat in both places was possible; the Incheon beach was only accessible six hours a day because of tide patterns; the approach to the port was narrow and could be protected by mines; and even if the landing were successful, Incheon resources might not be able to support the operation.

MacArthur reassured naysayers with all the positive points of the landing: The vast majority of North Korean troops were fighting in the Busan Perimeter; the enemy didn't expect an attack at Incheon, so thus would be unprepared; and the recapture of Seoul, only 20 miles from Incheon, would represent a significant moral and military victory. The plans for the Incheon landing moved forward.

After preliminary maneuvers, troops began landing on three different beaches on September 15, unloading vital equipment and overpowering any resistance. Although sporadic fighting took place, all three beaches were captured by the end of the day. On September 19, the troops captured Gimpo Airport, the largest airport in Korea and a strategic location essential to success. Next, they headed for Seoul.

While the landing on Incheon was swift and relatively easy, the march and recapture of Seoul was slow and laborious. Troops took 11 days to travel 20 miles, meeting resistance along the way and giving the North Koreans time to fortify the city. Once the soldiers entered the capital, they engaged in house-to-house fighting with desperate enemy forces.

U.N. troops that had fought in the Busan Perimeter routed the enemy ahead of them and connected up with U.S. X Corps. Together, they defeated the last fighting North Korean soldiers in South Korea. ■

MG Oliver P. Smith, CG, 1st Marine Division (left), receives a friendly gesture from Gen. MacArthur (right), who has just presented General Smith with the Silver Star. (September 21, 1950, photo by the NARA)

General MacArthur

Born and raised in a military family (his father, Arthur MacArthur, Jr., was the highest-ranking Army officer at one time), General Douglas MacArthur was destined to serve in the military. MacArthur became one of only a handful of men to earn the rank of a five-star general, and is the only man to serve as a field marshal in the Philippine Army. He played a prominent role in the Pacific Theater during World War II, and led the United Nations Command during the Korean War until Pres. Truman relieved him of duty.

Born on January 26, 1880, at the Little Rock Barracks in Arkansas, MacArthur's early childhood was spent on western frontier outposts where his father was stationed. "It was here I learned to ride and shoot even before I could read or write," he noted.

MacArthur graduated as valedictorian of the West Point Class of 1903 before serving as a junior officer in the years leading up to World War I. During that time, he was stationed in the Philippines, served as an aide to his father in the Far East, and participated in the American occupation of Veracruz, Mexico. In World War I, he formed the remarkable Rainbow Division by combining troops from many National Guard units. He rose in rank to become division chief of staff, brigadier general, and divisional commander. Following the war, he led significant reforms while superintendent at West Point, held two commands in the Philippines, and led the 1928 American Olympic Committee. His service during World War II established him as the ideal officer to head the military efforts in the Korean War.

In April 1951, after he was relieved of duty, MacArthur returned to the United States, where he was welcomed as a hero, although he continued to be a critic of Pres. Truman's policy in Korea.

Gen. MacArthur lived the remainder of his years quietly, serving as chairman of Remington Rand, a maker of electrical equipment and business machines, and spending time with his family. He died at age 84 on April 5, 1964, at Walter Reed Army Hospital in Washington, D.C., and was buried at the MacArthur Memorial in Norfolk, Virginia. ■

TOP: **Four LSTs unload on the beach at Incheon as U.S. Marines gather equipment to move rapidly inland on September 15, 1950. Landing ships were stuck in the deep mud flats between one high tide and the next.** (September 15, 1950, photo by C. K. Rose, Navy)
BOTTOM: **South Korean troops raise their national flag in front of the Capitol after retaking Seoul from the North Koreans on September 27, 1950.**
(Photo from Yonhap News)

SECTION II

FIGHTING BACK

A Navy AD-3 divebomber pulls out of a dive after dropping a 2,000-pound bomb on the North Korean side of a bridge over the Yalu River at Sinuiju, into Manchuria. (November 15, 1950, photo by the Navy)

CHAPTER 5

North to the Yalu

The Korean War was fought, in large part, on the ground by infantry soldiers. While air and water resources were critical to the successful outcome of the war, much of the action took place in the trenches. This was nowhere more evident than in the push northward following the U.N. victories in the Busan Perimeter, the landing at Incheon, and the recapture of Seoul.

After these successes, President Truman sent General MacArthur a top-secret National Security memorandum, authorizing him to unite all of Korea under Syngman Rhee, if possible. The orders to march, however, came with one strict limitation: the northward offensive should only continue if China and the Soviet Union remained out of the war.

By October 1, the U.N. Command repelled the Korean People's Army northwards, past the 38th Parallel; the Republic of Korea's forces crossed after them into North Korea, followed a few days later by the U.N. Command forces. North Korea is divided down its center by the rugged Taebaeksan Mountains, making it necessary for the U.S. 8th Army, with its U.N. units attached, to drive northward on the west side of the range, while U.S. X Corps landed on the east coast and advanced northward on a separate front to the east side of the range.

The ground troops that drove up western Korea captured a string of cities, including Pyongyang, the North Korean capital, on October 19. Their destination was north to the Yalu River, the boundary between China and North Korea—despite warnings by China that they would attack if U.S. troops crossed the 38th Parallel.

At this point, General MacArthur assured President Truman and others that China had no real intention of intervening and was only making empty threats. Even if the country did decide to get involved, MacArthur insisted, their forces would be spotted and destroyed by air power.

In truth, however, although MacArthur didn't realize it, the People's Republic of China had already entered the war. Chinese reinforcements had begun marching toward the U.N. forces' location and, under cover of night, were positioning themselves for their first offensive move, which would once again turn the tide of the war.

The first fighting between the U.N. and Chinese troops took place at the Battle of Unsan on November 1, when thousands of Chinese began a surprise attack. Unprepared, U.N. forces retreated. However, this initial victory caused Soviet Premier Joseph Stalin to change his mind, and the Soviets also became more involved in the war, providing crucial air cover, equipment, and other supplies to the reinvigorated North Korean military. ■

Marines searching a shattered building in a methodical quest for North Korean stragglers and suicide snipers. (Photo by the NARA)

Dropping Down a Message

Solomon Jamerson, who was awarded a Silver Star for his service in the Army, recalls how it felt after the success at the Busan Perimeter and Incheon landing: "We had been talking about having to be evacuated from the peninsula, and with the Incheon invasion coming in the middle of September, that took away all the possibility of them pushing us off… We started moving forward, and morale was very high at that point."

Jamerson served as an aerial observer and recalls flying over enemy territory one day, noticing someone who looked like a farmer walking along a road. Something didn't seem right, however, and Jamerson sensed an ambush targeting advancing ground troops. Sure enough, a shot burst out of a nearby house. Jamerson knew he had to let the troops below know. "We had a little message pad in the aircraft, and you wrote your little message out and put it into a bag with a heavy metal weight on it, and you go and drop it off," he said. "That was our only communication with the infantry… After we dropped this message down to the commander, we went back up and continued directing fire on all the various locations."

It was getting close to nightfall, and there were no landing lights at the base, so Jamerson and his pilot were told to return. "[But] the pilot said, 'I'll stay with you as long as you think you can do some good up here.' So we continued firing and then finally had to take off to the base."

Initially, the pilot was threatened with a court martial because of his refusal to return to the base, "[But] the commander down on the ground was so pleased with our actions, he put us both up for the Silver Star," Jamerson noted. "I've never had the opportunity to find out who that commander was, but I would have loved to have heard what he thought about the activity we had that afternoon." ■

Men of the 1st Cavalry Division fighting in a train station in Pyongyang, Korea. (October 19, 1950, photo by the NARA)

"That All Changed When the Chinese Came"

Army Private Donald Byers arrived in Korea as a replacement soldier for the Incheon landing. With the success there, and the rapid retreat of the North Korean troops, many soldiers thought the war was over. "I was just infantry," he notes, "an ammo-bearer to the machine gun squad. Everything was great; the war was all over."

"That all changed when the Chinese came in," Byers recalled. "It was pretty horrific. You wake up in the middle of the night, and all hell was breaking loose. There were bombs going off and artillery, and bullets going off right in front of our noses. We set up our machine guns and were firing across the river and into the mountains—that's where the flashes of gunfire were coming from. It was bitter cold. Thirty-eight degrees below zero, and no place to go to get out of it."

Byers and his buddies fought the Chinese bitterly for weeks in seesaw battles back and forth. "This was war," he says, "not much but cleaning weapons and shooting what you could." Typically, the shooting took place at night, and during the day, the soldiers would pull back and rest. Near the end, that changed. "[The Chinese] just kept on charging, during the daytime and nighttime too. We were firing our machine guns, using up all the ammo we could, and killing many of them."

"We were totally surrounded," Byers said, "and we didn't think we were getting out of there. We just thought, 'Kill as many of them as we can because we're going to die anyway.' Being a Pfc., I didn't get to talk to commanders; we didn't know anything about the strategy of the war. All I saw was my platoon sergeant, and he just said, 'Keep firing.'"

"On the third day, when the task force came in with the tanks and broke us out of the encirclement, the Chinese just melted away into the mountains from there, so we knew that we were safe, that we were going to get out." ■

The railway station and plaza at Daejon, Korea, are scenes of great activity as refugees flee south from the Communist invaders and U.S. and Korean troops move north to the battle front. (July 6, 1950, photo by the NARA)

Convoy of DUKWs (6-wheel-drive amphibious trucks), carrying American and ROK Marines, move to Han River in offensive launched against the North Korean forces in that area. (September 20, 1950, photo by the NARA)

China Joins the War

After their successive losses and frantic retreat in September, the North Koreans knew they could not continue the war without help. North Korean President Kim Il Sung had been in constant contact with both Chinese and Soviet leaders, pleading for essential equipment and especially more soldiers. Although Stalin said the Soviets would not directly intervene, China reluctantly responded to the request, feeling obligated because tens of thousands of North Koreans had fought in China's recent civil war.

Although Gen. MacArthur felt certain any movement by the Chinese would be spotted by U.N. aerial reconnaissance efforts, Chinese People's Volunteer Army began their involvement in the war undetected. They accomplished this by only moving at night and hiding under camouflage by sunrise. During daylight or while marching, soldiers remained motionless if an aircraft appeared; violators who exposed themselves to possible enemy observation were shot. Under these conditions, a three-division army marched 286 miles to the combat zone in 19 days, resulting in a surprise attack against the U.N. Corps. ■

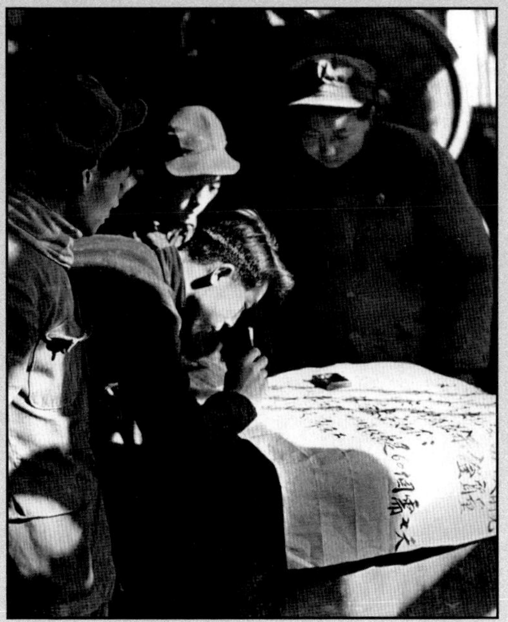

ABOVE: 1950s, China: Volunteers from a North China machine factory sign up for military service in North Korea. (Photo by the NARA)

B-29s of the U.S. Air Force drop their 500-pound bombs on a strategic target in North Korea. These planes devastated enemy North Korea supply lines, industrial areas, and troop concentrations with their precision bombing. (Photo courtesy of Air and Space Museum)

CHAPTER 6

MiG Alley and Air Power

The Korean War marked several firsts for the U.S. Air Force. It was the first war that extensively used jets and helicopters, the first major war against an agricultural nation, and the first war during the nuclear era. It also marked a last war where propeller aircraft was predominantly used.

Initially, the war didn't start out with significant air presence. The North Korean forces depended on a small force of aging, propeller-driven, Soviet aircraft left over from World War II, and Pres. Truman wanted the U.S. Air Force strength focused in Europe, to serve as a strong warning against the Soviet Union to avoid any action.

Although the Air Force, much like the other branches of the military, was not expecting war in Korea, it responded quickly. The older B-29 bombers did their job well, wreaking havoc on North Korean communication centers, military installations, and transportation networks, as well as slowing down the advance of the enemy. In addition, the Air Force transported much-needed troops and equipment from Japan to Korea, evacuated American citizens, and provided important intelligence through aerial reconnaissance.

During the war, the northwest portion of North Korea, where the Yalu River empties into the Yellow Sea, became a key location for the Air Force. Dubbed "MiG Alley" because the Soviet MiGs made their first appearance there, it was the site of numerous dogfights, and is considered the birthplace of jet-to-jet combat.

Primarily an agricultural nation, Korea offered few industrial or military targets, so the Air Force's traditional strategic bombing was pointless. Instead, the Air Force focused on raids on supply routes, bombing tanks, moving troops, and flying missions that required close coordination with land or sea forces.

Airlifts also played a key role in the war, especially during the first year when ground troops were often separated during rapid advances or retreats and had to be rescued. In addition, air services were essential for delivering food, ammunition, medical supplies, and mail. Cargo planes made much larger deliveries, including jeeps, big guns, and even a bridge; they also evacuated wounded soldiers to Japan.

With assistance from the U.S. Navy and Marines, the U.S. Air Force maintained air superiority throughout the war and played a significant role in ending the war. ■

TOP: Lt. R. P. Yeatman, flying from USS *Bonhomme Richard,* rockets and bombs a bridge below to disrupt supply lines from North Korea. (November 1952, photo by the NARA)
BOTTOM: Paratroopers of the 187th RCT [Regimental Combat Team] float earthward from C-119s to cut off retreating enemy units south of Munsan, Korea. (March 23, 1951, photo by Cpl. P. T. Turner, Army)
OPPOSITE PAGE RIGHT: Navy Sky Raiders from USS *Valley Forge* fire 5-inch wing rockets at North Korean field positions. (October 24, 1950, photo by Petty Officer 3rd Class Burke, Navy)

S
S
508
S
S

Tons from the air

On August 12, 1950, the USAF dropped 625 tons of bombs on North Korea; two weeks later, the daily tonnage increased to some 800 tons. U.S. warplanes dropped more napalm and bombs on North Korea than they did during the whole Pacific campaign of World War II. As a result, 18 North Korean cities were more than 50% destroyed. The war's highest-ranking American POW, U.S. Major General William F. Dean, reported that most of the North Korean cities and villages he saw were either ruins or snow-covered wastelands. ■

TOP LEFT: F4Us (Corsairs) returning from a combat mission over North Korea circle USS *Boxer* as they wait for planes in the next strike to be launched from her flight deck—a helicopter hovers above the ship. (September 4, 1951, photo by the Navy)

TOP RIGHT: View of F-86 airplanes on the flight line getting ready for combat. (June 1951, Photo by the NARA)

RIGHT: Air Rescue helicopter about to land at an advanced air station in Korea. During one operation, 12 Sikorsky S-55 helicopters moved a battalion of 1,000 Marines more than 16 miles in four hours. (Photo by the NARA)

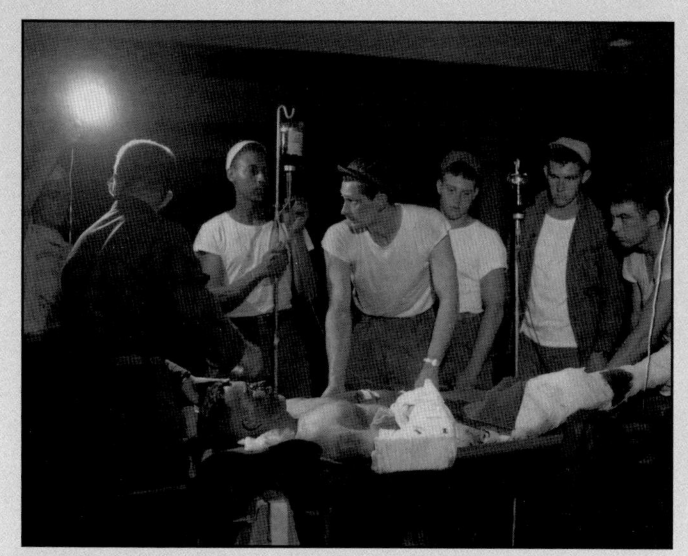

ABOVE: A seriously wounded soldier of the 116th Engineers, prior to his operation at the 121st Evacuation Hospital, in Yeongdeung-po. (August 17, 1951, photo by G. Dimitri Boria, photo by the Army)

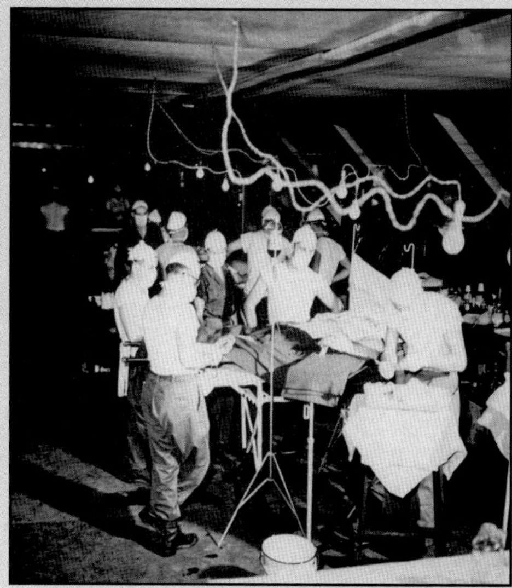

ABOVE: Surgery is performed on a wounded soldier at the 8209th Mobile Army Surgical Hospital, 20 miles from the front lines. (Photo by the Army)

MASH Units

Helicopters made their first war-time appearances during the Korean War, specializing in transporting soldiers in trouble. They were especially effective in evacuating wounded men from the front lines and flying them to nearby Mobile Army Surgical Hospitals, or MASH units, which also made their first war-time appearances during the Korean War.

The purpose of MASH units was to place experienced medical personnel closer to the front so that they could treat the wounded sooner and more successfully. Casualties were first evaluated and treated at the front line, then moved to a battalion aid station, and finally transferred to the MASH. During the Korean War, a seriously wounded soldier who made it to a MASH unit alive had a greater than 97 percent chance of survival once he received treatment.

"The creation of MASH units was a really fine move on the part of the military," observes Dr. Dale Drake, an anesthesiologist who spent 16 months serving in a MASH. "It was a place where more definitive surgery could be performed with qualified workers, certified surgeons, and anesthesiologists and so on, and yet it was only a few miles from the action of war. At night, at least from the tent I was in, you could see the flash of artillery, and it seemed like it was not very far away, perhaps seven or eight miles."

Drake and his wife, Cathy, met while they were both serving in a MASH unit during the war; Cathy was a nurse. They both worked with Dr. H. Richard Hornberger, who wrote several books after he returned from the Korean War under the pseudonym of Richard Hooker. He collaborated with W. C. Heinz on one of those books—*MASH:* A Novel about Three Army Doctors—which inspired the well-known movie and TV series of the same name.

After the war, the Drakes visited Hornberger one night, and they sat up well past midnight, reminiscing about their MASH days. "Dr. Hornberger had a man there who was a writer, and we didn't know that," Cathy recalls. "He was just visiting. So we told all these kinds of stories, and laughing and everything. I guess maybe a year later or six months later, we get the first book of his, and it said, 'Born the night you blew in.'"

Although the book was written based, at least in part, on some of the memories the Drakes shared that night, life in a MASH unit was definitely different than shown in the movies. "I think that's all for consumption by the TV public," Drake observes. "The real thing, it should have been serious, and it was serious. As far as social life at the MASH, I'm not saying it was non-existent, but there wasn't much to offer, really." ■

Marines of the 5th and 7th Regiments, who stood up to a surprise onslaught by three Chinese divisions, withstand a winter storm.
(Photo by Sgt. Frank C. Kerr, Marine Corps)

CHAPTER 7

A Brutal Winter

Winter began warmly enough, with the U.S. 7th Division feasting on hot turkey dinners upon reaching the south bank of the Yalu River on the East side of North Korea only three days before Thanksgiving. The campaign to win the war by Christmas seemed imminent. Little did the 7th Division know they were in for the most brutal winter they could imagine. The massive influx of Chinese forces over the North Korean border surprised U.N. forces, shattering any ideas of Christmas at home. Instead, U.N. troops were ordered to fall back on all fronts. Troops had to give up ground they had gained valiantly, as the biggest evacuation in U.S. military history began.

While the U.N. worked to arrange a cease-fire, U.N. forces fought hard to maintain critical positions as troops north of them retreated. Bridge sections were dropped by parachute to rebuild critical bridges needed for escape. Marines carved an airstrip in the ice so that wounded, too critical to move by land or sea, could be airlifted out. They held crucial mountain passes for days at a time in detestable conditions, they battled in weather 50 degrees below zero, and food and supplies were air-dropped to help them make their way.

Perhaps the most brutal battle of the winter was at the Changjin Reservoir (Chosin). The 1st Marine, 3rd Infantry, 7th Infantry Divisions, and the 41 Commando of the Royal Marines (UK) were trapped by the Chinese when the evacuation was ordered. General Oliver P. Smith exclaimed, "Retreat, hell! We're not retreating, we're just advancing in a different direction!" Retreat or advance, the soldiers went nowhere. Faced without an escape route, the Divisions fought hard managing to maintain their position against the onslaught of Chinese soldiers. When the road to escape finally opened, their retreat was slowed by continued battle.

Regimental Combat Team 31 of the 7th Infantry Division, later known as Task Force Faith, guarded the right flank of the Marine advance toward Mupyong-ni. Chinese forces east of the Reservoir nearly destroyed the task force team, and many of those who did manage to withdraw had to cross the frozen ice.

In all, the soldiers who managed to survive the battle at Changjin marched over 60 miles from the Changjin Reservoir to Hagaru-ri, through Hell Fire Valley to Koto-ri and then Hamhung before finally reaching Hungnam.

At Hungnam, the Navy Fleet commanded by Rear Admiral Doyle waited to finalize the evacuation. The 3rd Division continued to hold Chinese troops while U.N. forces boarded the Navy fleet. The evacuation totaled over 100,000 soldiers, 17,500 vehicles, 350,000 tons of equipment, and 91,000 Korean civilians. ■

"God was at the Helm"

Once the U.N. troops reached the port, the heroics continued. In just two weeks, the bulk of the U.N. forces in eastern North Korea, along with equipment, were moved from Hungnam in one of the most impressive evacuation movements by sea in U.S. military history. The numbers are staggering: A 193-ship armada assembled at Hungnam and evacuated approximately 100,000 military personnel, an estimated 17,500 vehicles, 350,000 tons of cargo, and 91,000 refugees. As the last U.N. vessel left, the port was destroyed so that enemy forces would be unable to use it. The operation was the largest sealift since World War II.

The SS *Meredith Victory*, a cargo freighter built for World War II, was part of this amazing evacuation. Designed to carry only 12 passengers, the ship evacuated more than 14,000 refugees during this single mission, earning it the title "Ship of Miracles."

While U.N. troops were being evacuated on ships, tens of thousands of war-weary civilians had also gathered at Hungnam, hoping to join the soldiers and flee from the approaching enemy. Captain Leonard LaRue couldn't say no, and he unloaded nearly all the weapons and supplies on the cargo freighter to transport as many refugees as possible. Koreans crammed into the five cargo holds and the entire main deck, even filling booms and elevators.

The passengers were "packed like sardines in a can" and most had to remain standing up, shoulder-to-shoulder, in freezing weather conditions and with very little food or water during the entire voyage.

The freighter departed for Busan on December 23, as gunfire from U.N. ships and explosives destroyed the port. "I think often of that voyage," LaRue said later. "I think of how such a small vessel was able to hold so many persons and surmount endless perils without harm to a soul. And, as I think, the clear, unmistakable message comes to me that on that Christmastide, in the bleak and bitter waters off the shores of Korea, God's own hand was at the helm of my ship." ■

TOP LEFT: Near Songsil-li, Korea, a tank of 6th Tank Battalion fires on enemy positions in support of the 19th RCT. (Photo by Pfc. Harry M. Schultz, Army)
TOP RIGHT: Chinese troops—wearing tennis sneakers, rags, and American foot gear—surrender to Charlie Company, 7th Marines, south of Koto-ri. (December 1950, NARA)
ABOVE: U.S. Marines move forward after effective close-air support flushes out the enemy from their hillside entrenchments. Billows of smoke rise skyward from the target area. (Photo by Cpl. McDonald, Marine Corps)

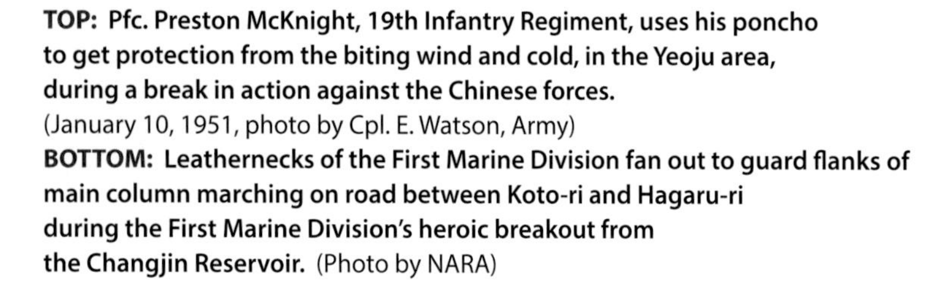

TOP: **Pfc. Preston McKnight, 19th Infantry Regiment, uses his poncho to get protection from the biting wind and cold, in the Yeoju area, during a break in action against the Chinese forces.**
(January 10, 1951, photo by Cpl. E. Watson, Army)
BOTTOM: **Leathernecks of the First Marine Division fan out to guard flanks of main column marching on road between Koto-ri and Hagaru-ri during the First Marine Division's heroic breakout from the Changjin Reservoir.** (Photo by NARA)

The Road Back—Astonished Marines of the 5th and 7th Regiments, who stood up to a surprise onslaught by three Chinese divisions, hear that they are to withdraw. In five days from November 28 to December 3, they fought back 15 miles through Chinese forces to Hagaru-ri, on the southern tip of the Changjin Reservoir, where they reorganized for the epic 40-mile fight down mountain trails to the sea. (Photo by the NARA)

TOP: Marines withdrawing back to the sea from Hagaru-ri, Korea. (November-December, 1950, photo by the NARA)
BOTTOM: First Division Leathernecks counter fire with fire when attacked by large Chinese formations during the Division's heroic breakout from the Changjin Reservoir. (December 7, 1950, photo by the NARA)

Nuclear Considerations

Setbacks prompted General MacArthur to consider using nuclear weapons against the Chinese or North Korean interiors, with the intention that radioactive fallout zones would interrupt the Chinese supply chains. Throughout the war, U.S. political and military leaders studied the possible use of nuclear weapons, and upon four separate occasions they gave this study serious attention. The answer was always the same: existing atomic bombs, carried by modified B-29s, would have little effect except for leveling cities. The one time that President Harry S. Truman suggested (in December 1950) that he was considering the nuclear option, the British led the Allies at the U.N. to stop such talk. ■

TOP: Marines withstand mountain winds and sub-zero temperatures as they move out from Koto-ri. (Photo by the NARA)
BOTTOM: First Marine Division takes to the road on the withdrawal from Koto-ri. (December 1950, photo by the NARA)

8

U.S. Marines fighting in the streets of Yeongdeungpo, south of Seoul, Korea. (September 20, 1950, photo by Lt. Robert L. Strickland and Cpl. John Romanowski, Army)

CHAPTER 8

Battles Over Seoul

As the new year dawned, the enemy struck southward in massive offensives, particularly across the Imjin River north of Seoul, as well as to the east along the west central front where they drove down onto Gapyeong and Chuncheon.

ROK units, taken by surprise, withdrew in great numbers. General Matthew B. Ridgway, who had taken command of U.S. Eighth Army following the death of Gen. Walton Walker in a driving accident in December, believed the Chinese might encircle the bulk of his army.

After conferring with his corps commanders, General Ridgway decided to strategically evacuate Seoul on January 3. Chinese commanders were astounded, but entered the capital the following day. From Seoul, the Chinese forces proceeded southward 30 miles to Osan on January 7, where American units held their advance in check. However, savage fighting went on in the mountains east of Seoul, where the Chinese offensive seemed unstoppable. Against recommendations of some of his senior staff, Ridgway ordered U.N. Commanders to cease rolling back and take to the offensive. The Chinese reacted with yet another offensive on February 11. U.N. Command forces on the west central front, east of Seoul, were driven down to the Wonju area.

One epic action of this era took place at Jipyeong-ni on February 13-15. The U.S. 23rd RCT commanded by Colonel Paul L. Freeman, Jr., with the 1st Ranger Company and an attached French Battalion, had been left in a salient position by the withdrawing U.N. forces. The Chinese moved on the 23rd RCT with 25,000 troops. After two days and nights, the enemy had suffered more than 5,000 casualties. Freeman's troops had held fast, having suffered 51 soldiers killed, 250 wounded, and 42 missing.

Ridgway launched the U.N. offensive Operation Killer on February 22, closely followed by Operation Ripper (March 7-April 4), with the objective of driving the enemy back beyond the 38th Parallel. Facing intense bombing and possible encirclement from the East, Chinese forces withdrew from Seoul. The U.S. 3rd ID and the ROK 1st ID liberated the capital city on March 14.

In April, President Truman relieved General MacArthur as Supreme Commander of the U.N. forces and Gen. Ridgway took over. The enemy launched another massive offensive in late April, but U.N. Command forces drove them back decisively.

By June 1951, both sides were forced to consider that a military victory might be too costly. Chinese and North Korean officials agreed to hold cease-fire discussions at the ancient capital of Kaesong on the north side of the Imjin River. An armistice to end the war seemed to be in the making. ■

TOP: U.N. troops continue advance: elements of the 27th Infantry Regiment, 25th Infantry Division pass by a burning house, as U.N. forces launch Task Force PUNCH against the Chinese forces eight miles southwest of Seoul. (February 7, 1951, photo by the NARA)
BOTTOM LEFT: A barricade held by men of the 1st Marine Division during the street fighting in Seoul. (Photo by the NARA)
BOTTOM RIGHT: Pfc. Albert Lumanais, 1st Battalion, 1st Regiment, 1st Marine Division, searches out remaining snipers as U.N. forces saturate Seoul. (Photo by NARA)

TOP: With a life-and-death fight for Seoul, a U.S. Marine spots a sniper and gets ready to return fire. (Photo by the NARA)
BOTTOM: U.N. Command Troops fighting in the streets of Seoul. (Photo by the NARA)

ABOVE: U.S. Maj. General Matthew B. Ridgway (left), appointed Supreme Allied Commander in 1951, with U.S. Gen. Mark W. Clark, Commander in Chief of the United Nations Command in Korea. (Photo by the NARA)

Truman Replaces MacArthur

On April 11, 1951, Pres. Truman relieved Gen. MacArthur of his command at the helm of the United Nations Command Forces. MacArthur had gained quite a bit of popularity in the United States, and even globally, for his leadership of the Allied Forces during World War II, and his dismissal during the Korean War remains controversial even today.

After World War II, MacArthur had been assigned to oversee the occupation of Japan, and was stationed in Asia when North Korea invaded South Korea. He seemed the natural choice to become the leader of the U.N. forces in the fight for freedom there. The highly successful Incheon landing had been his brainchild, a military move that many called genius. However, it was also the MacArthur-directed full-scale invasion of North Korea which followed that led to China joining the North Koreans.

Although the U.N. troops had been forced to withdraw from North Korea, they had made progress under Gen. Ridgway during the early months of 1951, and Pres. Truman saw an opportunity to suggest a negotiated peace. MacArthur knew the president's intention, but he publicly called for China to surrender instead and wrote a letter which was read on the floor of the U.S. House of Representatives on April 5, a letter critical of Truman's Europe-first policy and limited-war strategy in Korea. The letter ended with, "We must win. There is no substitute for victory."

The letter was the last straw. After months of private and public bickering, Pres. Truman called MacArthur home. In May and June of 1951, the Senate Armed Services Committee and the Senate Foreign Relations Committee held a joint inquiry into the circumstances surrounding MacArthur's relief, concluding that "the removal of General MacArthur was within the constitutional powers of the President, but the circumstances were a shock to national pride." ■

Pfc. Clarence Whitmore, voice radio operator, 24th Infantry Regiment, reads the latest news, near Sangju, Korea. (August 9, 1950, photo by Pfc. Charles Fabiszak, Army, USIA)

Fighting with the 2nd Infantry Division north of the Chongchon River, Sgt. 1st Class Major Cleveland, weapons squad leader, points out North Korean positions. (November 20, 1950, photo by Pfc. James Cox, Army)

Korean War Marks the End of Segregation

The Korean War was the beginning of desegregation in the military. Pres. Truman ordered the desegregation of the military with Executive Order 9981 on July 26, 1948; however, it would be some time before desegregation became a reality within the military. When the United States entered the Korean War in 1950, military units were still segregated, but as the military struggled with heavy losses, desegregation became necessary to maintain fully manned units. The desegregation of the military abroad was a catalyst for desegregation back home.

The U.S. Air Force had only been an independent military branch of service since 1947, so its segregation struggles were somewhat unique as it worked to establish policies and organization while also desegregating. The integration plan for the Air Force went before the Secretary of Defense on January 6, 1949. Logistically speaking, segregated flight units created problems, which may have contributed to the Air Force's swift actions to desegregate. Early in 1950, many touted the Air Force as the leader in military desegregation. Between 1949 and 1956, the percentage of black enlisted airmen rose from 5.1 percent to 10.4 percent, and black officers increased from 0.6 percent to 1.1 percent.

The U.S. Navy persisted during the Korean War in its efforts to attract African-Americans. The service dispatched recruiters to speak to black high school students. After a decline from 17,518 in 1949 to 14,842 in 1950, the number of African-American sailors reached 17,598 in 1951, and surpassed 24,000 by the end of fighting in July 1953. The number of blacks in the enlisted force had increased by 7,000 after establishment of the President's Committee on Equality of Treatment and Opportunity in the Armed Services Committee (also called the Fahy Committee after its chairman Charles Fahy) in 1949. Equally encouraging was the fact that some 2,700 blacks applied during 1949 to enter NROTC, either by competing for scholarships or joining units at colleges where they already were enrolled.

The Marine Corps also recognized the need for skilled

Men of the 24th Infantry Regiment move up to the firing line in Korea. (July 18, 1950, photo by the Army)

Pfc. Edward Wilson, 24th Infantry Regiment., wounded in the leg while engaged in action against the enemy forces near the front lines in Korea, waits to be evacuated to an aid station. (February 16, 1951, photo by Pfc. Charles Fabiszak, Army)

men of all races to efficiently fight the Korean War. As a result, on December 13, 1951, Marine Corps headquarters issued a memorandum directing subordinate commands to fill all billets with qualified Marines, regardless of race.

Three years after Pres. Truman had signed the Executive Order, the U.S. Army formally announced its plans to desegregate. Initial moves to desegregate were born out of a shortage of men in white units and a dramatic overstrength in black units. The disparity continued to widen until field commanders had no choice but to integrate; the results were noticeably favorable and caught the attention of many in command. Gen. Ridgway had personally observed the 24th Infantry's problems with segregation and felt strongly that desegregation was more efficient and more proper.

Certainly the Korean War was instrumental in the desegregation fight. Many African-American soldiers were heroes during the war, earning numerous medals, including Medals of Honor and Silver Stars, and achieving several individual honors. Roscoe Robinson, Jr., served as a platoon leader and rifle company commander in Korea, earning a Bronze Star and later becoming the first African-American to hold the rank of general. Ensign Jesse L. Brown was the first African-American aviator in the history of the U.S. Navy and he was posthumously awarded the Distinguished Flying Cross. Second Lieutenant Frank E. Petersen, Jr., became the Marine Corps' first African-American pilot and became the Corps' first flag officer. Captain Daniel "Chappie" James, Jr., received the Distinguished Flying Cross for his actions in Korea, and later became the first African-American to reach four-star rank in the military.

Segregation officially ended in the military in 1954, when the last segregated unit was disbanded. It had been nearly six years since Pres. Truman signed Executive Order 9981, and in that time nearly a quarter of a million black service members had been integrated into the nation's fighting force. But the fight for equality back home would take several more years. As soldiers returned from Korea, they often found their hometowns still completely segregated. ■

9

Infantrymen of the 27th Infantry Regiment, near Heartbreak Ridge, take advantage of cover and concealment in tunnel positions, 40 yards from the enemy.
(August 10, 1952, photo by Feldman, Army)

CHAPTER 9

Battles of the Punchbowl

Despite the Chinese agreement to discuss an armistice, Mao Zedong didn't want to end the war in defeat, and approved a plan to win limited victories through violent night attacks and infantry infiltration. So while officials worked to agree on peace, U.N. troops were forced to continue to fight.

Much of the fighting took place in the Punchbowl, a natural geologic bowl several miles across and encircled by steep mountains, creating a terrain that made planning and maneuvering particularly challenging. The Punchbowl battles were bitter and bloody, with thousands of lives lost.

One of those battles was Bloody Ridge. Beginning in mid-August, U.N. troops began moving to seize a series of hills where they believed enemy troops were hiding, and after a week of fierce fighting, succeeded in capturing most of the area.

The win was short-lived, however; North Korea counterattacked and the battle raged for 10 days. Ultimately the North Koreans abandoned the ridge, establishing position only 1,500 yards away on a seven-mile ridge that would soon earn the name Heartbreak Ridge.

This ridge was also dotted with hills, and a pattern of attack quickly emerged: U.N. troops began bombarding a hill with aircraft, tanks, and artillery fire before infantry soldiers would clamber up the rocky slopes, taking each enemy bunker by hand-to-hand combat. The victors, however, were exhausted, short on supplies, and ill-equipped for the inevitable counterattack, launched by fresh North Korean troops.

This pattern continued for two weeks, before a new strategy was planned to cut off reinforcements. This plan required tanks, so U.N. forces were deployed to build a road for the tanks. These soldiers worked under enemy fire much of the time but accomplished the task, and tanks began rolling in.

Coincidentally, the U.N. attack started just as a Chinese division was moving toward Heartbreak to relieve the North Koreans. They got a glimpse of the strategy, and when fighting started, anti-tank trenches and guns were in place. The enemy killed soldiers and destroyed tanks, before U.N. Command forces cut off supply roads and won the hills along Heartbreak Ridge. After 30 days of costly battle, U.N. forces stood victorious.

These were just two of the many Punchbowl battles, each one important in its own way as U.N. officials negotiated an end to a war that had gone on far too long with no substantial gains for either side. ■

ABOVE: Men from U.S. 25th Infantry Division build a new CP bunker on Hill 931, Heartbreak Ridge, Korea. (October 9, 1952, NARA)

Only One Way To Go—Forward

Donald M. Cohen was one of the ground soldiers in a unit that fought at both Bloody Ridge and Heartbreak Ridge. "They wanted that hill taken because if the Chinese or North Koreans had it, they would be able to see everything south," he explains, "and we wanted to take that from them. But there were so many, you know, they could outnumber you, 20 to one. And they were very good in artillery. They could put it right in your back pocket. We just kept firing, shooting and shooting... You stay close to the ground and keep moving; you got to keep moving. If you stop, you're gonna get killed. It's that simple... The first time I saw somebody killed, it was a hell of a shock. I'd never seen anything like this before. In the movies, you know, but you eat your candy bar and popcorn and go home. But there was no John Wayne, no band music playing, all you hear is explosions and bullets firing over your head and grenades going off. So you're confused, but somewhere down the line, the training takes over, it's self-preservation. You're not gonna run back, no way, 'cause they'll shoot you in the back. So there's only one way to go—forward."

TOP: Men of the Heavy Mortar Co., 7th Infantry Regiment, cook a much-needed hot meal in their foxhole in the Gagyedong area, Korea. (December 7, 1950, photo by Pfc. Donald Dunbar, Army)

BOTTOM: A wounded U.S. Marine awaiting transport back to a field hospital after receiving First Aid at the battlefront. Shell and mortar fire inflicted over 1,000 casualties per month on the 1st U.S. Marine Division in the winter of 1952-53. (Photo by the NARA)

TOP LEFT: Men of the 5th Marine Regiment, 1st Marine Division hit the dirt when the Chinese opened up fire on them from a ridge 500 yards north of the Marine line heading towards Jaun-ni, Korea. (Photo by the NARA)
TOP RIGHT: U.S. Soldiers securing the top of the hill known as "Bloody Ridge" near Yanggu in Gangwon-do, Republic of Korea, following the battle of "Heartbreak Ridge." (Photo by the NARA)
BOTTOM: Sniper hunt: a U.S. Marine patrol closes in on a Korean hut in search of a sniper offering stubborn resistance in front-line action. (Photo by the NARA)

SECTION III

WAITING FOR A RESOLUTION

10

Pfc. Roman Prauty, a gunner with 31st RCT (crouching foreground), with the assistance of his gun crew, fires a 75mm recoilless rifle, near Oetlook-tong, Korea, in support of infantry units directly across the valley.
(June 9, 1951, photo by Peterson, Army)

CHAPTER 10

The Iron Triangle

Despite heavy losses in the Punchbowl, Chinese leaders felt their "active defense" had worked. U.N. forces had given up major offensive operations and, in fact, would spend the rest of the war concentrating on defensive strategies. However, this change had more to do with the U.S. public's growing protest against limited-objective battles than with being intimidated by Chinese manpower.

The armistice talks moved to Panmunjeom in October 1951, and both sides agreed on most issues, including the creation of a demilitarized zone and enforcement of the armistice after the shooting stopped.

What neither side could agree on, however, was the handling of prisoners of war. For almost two years, officials would try to reach a consensus on what to do with the tens of thousands of Communist Chinese and North Koreans who, the American officials correctly supposed, would rather not return to their homelands.

And during those two years, despite the fact that both sides knew peace was in sight, the fighting raged on. In one of the few U.N. offensives during this time, several attacks were made in the Iron Triangle, a key transportation and communication area approximately 60 miles north of Seoul, and the most direct route to the Republic of Korea capital.

Key battles for ownership in the area included the Battles of White Horse, Triangle Hill, and Pork Chop Hill. During 10 days of fighting at the Battle of White Horse Hill, the hill changed hands 24 times before the U.N. finally conquered it. The resulting destruction was so complete that the hill looked like a threadbare white horse, hence its name. Apart from American tank, artillery, and air support, the battle was fought completely between ROK and Chinese troops.

The battle of Triangle Hill followed close behind; this conflict was bitter and bloody but not nearly as successful. After more than a month of repeated attempts to capture the area, including nearby Sniper Ridge, escalating numbers of casualties forced U.N. leaders to call off attacks, and the Chinese regained original ownership.

The Battle of Pork Chop Hill was a two-part battle; U.N. troops won the first battle in April when the Chinese withdrew after only two days of fighting. A second battle in July, however, involving many more soldiers on both sides, resulted in a five-day battle that ultimately ended with a U.N. withdrawal. ■

Holding Outpost Harry

A pivotal battle in the Iron Triangle took place at Outpost Harry, with more than 88,000 rounds of artillery fired by the Chinese alone. A strategic location dearly desired by the Chinese, Outpost Harry blocked the Chinese view down the valley and shielded a portion of the area from direct enemy fire. Even more importantly, U.N. officials felt that losing that outpost may cause the Chinese to continue fighting instead of reaching an armistice agreement, which they felt was fairly close. U.N. troops were told to "hold Outpost Harry at all costs."

The outpost was only guarded by a single company of either American or Greek soldiers, so the Communist enemy expected an easy victory. Over the period of eight days, some 13,000 Chinese soldiers flooded the area, while five companies (four American and one Greek) took turns defending the outpost. Most of the fighting took place during the dark of night, with daylight hours interrupted sporadically by enemy gunfire but spent primarily removing the dead, treating the wounded, and repairing and fortifying the area.

Ultimately, the U.N. troops did hold on, although the cost in lives was high. The five rifle companies received the distinctive Distinguished Unit Citation for their heroic efforts during this battle, the first time in history that five companies shared that honor. ■

TOP LEFT: Wounded soldier helped to safety during the conflict for Pork Chop Hill. (Photo by the NARA)
TOP RIGHT: ROK Soldier keeps a lookout on White Horse Mountain, as the division recaptures the Mountain 24 times in 10 days. (Photo by the NARA)
MIDDLE: Medical Corpsmen assist wounded infantrymen following the fight for Triangle Hill (No. 598).
(Photo by the NARA)
BOTTOM: U.S. Marines of the First Marine Division Reconnaissance Company make the first helicopter invasion on Hill 812, to relieve the ROK Eighth Division.
(Photo by T. G. Donegan, Navy)

TOP: Supply warehouses and dock facilities at an east coast port feel the destruction of para-demolition bombs dropped by the Fifth Air Force's B-26 Invader light bombers. (Photo by the USIA)
MIDDLE: The USS *Missouri* fires 16-inch shell into enemy lines. (Photo by the NARA)
BOTTOM: Chinese prisoners captured during conflict for Pork Chop Hill, Korea, are evacuated to rear areas by Infantrymen of the 31st Infantry Regiment, 7th U.S. Infantry Division. (April 17, 1953, photo by the NARA)

The Battlefront Is a Terrible Place

Kenneth F. Gibson, a sergeant in the Marines, recalls the last few weeks of the war: "We were confined to South Korea. I got there after they had already crossed the 38th Parallel and went up to the Manchurian border, and that's when the Chinese came into the war. And we withdrew all the way down into Busan before we were reorganized and fought our way back to the 38th Parallel, where the war came to a final conclusion. It was a devastated place, and Korean civilians who had gone as far south as they could to escape the battle were living in a mess, just like any other war-torn country. We fought in combat; my particular unit was a small artillery platoon, and I had 29 men in the platoon. We had so many Chinese and North Korean soldiers against us that we were almost wiped out. We did not retreat. We stayed. And we were replaced at one time with another Marine company, and we went back and rebuilt, got more men, and went at it again. The battlefront is a terrible place... sometimes it's hard to talk about those experiences." ■

Scrimmage Line

Lou Sardina was a private working on a tank who fought in the last battles of the war. "It was rough going," he says, "because it was nothing but taking hills and killing. That was all we did. That first morning I can still remember just as vividly as it if happened yesterday... The Chinese were shooting at us so the bullets were going above our heads; you could hear them."

"The machine gunners go over your head," Sardina continued, "but as soon as you get too close, the machine gunners have to stop or they'll shoot you, so now it's between you and the enemy. And that's called a scrimmage line... We were the first ones up, and it was tough going. Guys got killed. What you're doing is you're going uphill and they're shooting and throwing grenades down at you, but they can't shoot down straight if the hill is up straight, but once you get to that point... you're not more than fifty, twenty yards. And now you're just going after them. This is it; this is combat. That's what it is. And up you go. And they're all shooting and you're going up just like it was the Civil War, like taking charge... You're in another zone. The blood and adrenaline is ripe, and you don't even know... Climbing up those hills, if you did it on your own you'd be tired, but you're not. But... after you take the hill, that's when you shake. The adrenaline starts, and you just shake. It was 10 days that we did that. One day just ran into the next. You didn't know if it was Sunday or Monday or Tuesday or Wednesday. It was just up and down one hill and the others, and the whole division was pushing... and that's where the line was stagnant until the armistice." ■

11

The families of the returning POWs waving and greeting USS *General Nelson M. Walker* as it docks at Fort Mason, California.
(August 23, 1953, photo by Pfc. Brink, Army)

CHAPTER 11

The Fighting Ends

Armistice negotiations had barely progressed for months, but in March 1953, Joseph Stalin died. Support for the war quickly eroded in the Soviet Union, with the Soviets voting to end the war.

Chairman Mao knew he could not continue the war without Soviet assistance, and negotiations began to move much faster, with the Chinese ultimately agreeing to voluntary repatriation. That meant that POWs who wanted to return to their homelands would be released immediately, while those wanting to stay would be transferred to the care of a neutral nation for screening. The Chinese and North Koreans also agreed to the exchange of sick and disabled POWs.

Republic of Korea leader Syngman Rhee became the obstacle now. He had never publicly given up the charge to march north and unify, and in private he had alluded to the fact that he would only agree to an armistice if the United States agreed to a mutual security alliance and pledged $1 billion in economic aid.

Ultimately, however, Rhee was persuaded to accept the outlined stipulations, although he never signed it. On July 27, 1953, the Armistice Agreement was officially signed by representatives from the United Nations, China, and North Korea. Interestingly, although the agreement called for continued peace talks, no peace treaty was ever signed.

The border between the Republic of Korea and North Korea returned to roughly where it had been before war broke out—near the 38th Parallel, dipping slightly below the Parallel in the west but extending far beyond the Parallel in the east. The two Koreas were separated by the Demilitarized Zone (DMZ), a 2.5 mile-wide buffer zone that extends for about 160 miles across Korea.

Adherence to the Armistice Agreement is monitored by members of the Neutral Nations Supervisory Commission. In addition, large numbers of troops from the two Koreas are stationed on both sides of the line and along the coastline and on outlying islands, making it the most heavily militarized border in the world.

The armistice agreement outlines exactly how many military personnel and what kind of weapons are allowed in the DMZ, as well as the patrolling action allowed by soldiers from both sides. Since the signing of the Armistice, sporadic outbreaks of violence and North Korean hostilities have resulted in the deaths of more than 500 Republic of Korean soldiers and 50 U.S. soldiers along the DMZ. ■

TOP: Members of the 223rd Infantry Regiment, 40th U.S. Infantry Division, listen to the announcement of the signing of the truce in their bunker on Heartbreak Ridge, Korea.
(Photo by the NARA)

BOTTOM: Gen. W. K. Harrison, Jr., signs armistice ending 3-year Korean War. Gen. Harrison, left table, and North Korean Gen. Nam Il, right table, sign documents.
(July 27, 1953, photo by F. Kazukaitis, Navy)

Col. James Murray, Jr., USMC, and Col. Chang Chun San, of the Korean People's Army, initial maps showing the north and south boundaries of the demarcation zone, during the Panmunjeom cease-fire talks. (October 11, 1951, photo by F. Kazukaitis, Navy)

ABOVE: 1st Lt. Alvin Anderson, one of the many repatriated POWs to return home aboard USNS *Marine Phoenix*, embracing his mother and sister. (September 14, 1953, photo by Herb Weiss, Army)

LEFT: Repatriated POW Capt. Fredrick Smith greets his father. (Photo by Herb Weiss, Army)

12

Korean POWs being exchanged at Panmunjeom, April 22, 1953.
(Photo from Yonhap News)

CHAPTER 12

Operation Glory

After the war ended, a priority for the United States was bringing home her soldiers: those who had survived the war, as well as those who had died.

The handling of the POWs had been the main sticking point all along in the armistice discussions. Initially, the U.N. representatives had assumed that all POWs would be returned to their homelands, but Republic of Korea officials vehemently opposed involuntary repatriation, knowing that thousands of POWs being held in the ROK were actually Southern citizens who had been forced to fight for North Korea. On the other hand, Chinese leaders knew that some of their soldiers, who had also been forced into the army, would refuse to return unless repatriation were mandatory.

Ultimately, all involved agreed that POWs be given a choice, and it was under that guideline that repatriation began. From August 5 to September 6, a total of 75,823 Communist soldiers and civilians (all but 5,640 of them Koreans) returned to their chosen country, while 7,862 Republic of Korea soldiers, 3,597 U.S. servicemen, and 1,377 persons of other nationalities (including some civilians) returned to U.N. control.

The handling of those who chose not to return to their homeland turned into a nightmare as they underwent the screening process. Communist agents had infiltrated the POW camps and made life miserable for those who were trying their best to accomplish their assignment. In February 1954, the Neutral Nations Repatriation Commission gave up the screening process.

After the war, Operation Glory called for the fighting countries to exchange their dead. From July to November 1954, the remains of 4,167 U.S. Army and Marine Corps dead were exchanged for 13,528 North Korean and Chinese dead; in addition, 546 civilians who had died in U.N. POW camps were delivered to the Republic of Korea government.

Many soldiers remain unaccounted for. U.S. Department of Defense Prisoner of War/Missing Personnel Office records show that China and North Korea transmitted a list of 1,394 names; only 858 were correct. In addition, from 4,167 containers of remains returned from the two countries, forensic examination identified 4,219 individuals, with 2,944 identified as Americans.

The quest to find the remaining MIAs continues. In the early 1990s, North Korea excavated and returned more than 208 sets of remains, but very few have been identified. From 2001 to 2005, more remains were recovered from the Changjin Battle site, and around 220 were recovered near the Chinese border between 1996 and 2006. ■

They Knew We Would Take a Stand

Arden Rowley arrived in Korea on July 31, 1950, and four months later was captured by the North Koreans. "By the end of November, the U.N. forces were approaching the Yalu River," he explains. "We had been left as a rear guard force to fight the Communists off so the other divisions could set up new lines of defense," Rowley recalls. "When it came time to withdraw, we were just completely overwhelmed with Communist forces. On the first of December, some 320 men were taken prisoners of war, and I was among those.

"When our battalion commander saw that the situation was hopeless and knew that capture was imminent, he ordered as much of the equipment destroyed as possible. I threw my grenades on the engine blocks of the truck and set fire to the tires on the truck, soaked with gasoline, and then he said, 'We don't want the colors to be captured by the enemy, to be used as a trophy of war.' So he ordered the colors be burned on the night of November 30.

"[My] battalion served in Korea until 2004, and those who served after the war took up this tradition of ceremonially burning the colors. I was able to attend three of those, and it was really an honor to be there, especially in 2000 when it was the 50th anniversary of the event. A tremendous feeling of nostalgia, a feeling of camaraderie. There were at least a dozen of us Korean War veterans who were there on November 30 who attended.

"It's quite a feeling to think back those 50 years earlier, when so many men gave their lives, sacrificed in many ways, being wounded, killed, taken prisoners of war. But I've talked to a lot of Korean veterans and [they] know in their hearts that every sacrifice we made over there was worth it because we stopped Communist aggression in that part of the world. And I think that probably precluded them from trying it anywhere else in the world at that time; they knew we would take a stand against that kind of aggression." ■

TOP: North Korean prisoners, taken by the U.S. Marines in a foothills fight, file across a rice paddy. (1950, photo by the Marine Corps).
MIDDLE: A U.S. Marine tank passes by a line of POWs down a village street.
(September 26, 1950, photo by Staff Sgt. John Babyak, Jr., Marine Corps)
BOTTOM: U.S. Marines guarding three captured North Koreans. (1950, photo by Sgt. W. M. Compton, Marine Corps)

ABOVE: Men of the 1st Marine Division capture Chinese soldiers during fighting on the central Korean front.
(March 2, 1951, photo by Pfc. C. T. Wehner, Marine Corps)

SECTION IV

A NEW BEGINNING

13

World Trade Center Seoul and COEX Convention & Exhibition Center, the heart of Korea's international commerce. (Photo by Seoul Selection)

CHAPTER 13

From Rubble to Powerhouse

When the guns finally fell silent on July 27, 1953, Korea was a shattered nation. Three years of war had leveled its cities, devastated its infrastructure and left countless dead or wounded. The Armistice left the nation divided just as it had been before the war, separating families, severing road and rail traffic electrical transmission lines, and presenting the new nation with a permanent threat from the north. Seoul had changed hands four times during the war, leaving much of the capital in ruins, but with refugees pouring in from the North, the city was reduced to a giant refugee camp. Dependent on foreign assistance, South Korea was, economically speaking, in worse shape than many sub-Saharan African states. The future looked bleak.

Then a miracle happened. Thanks to the indomitable will of the Korean people to not only survive, but prosper. Korea pulled itself up from its own bootstraps with wise decision-making, sacrifice, and plenty of hard work. Within just half a century, Korea had transformed itself into a commercial, political and cultural powerhouse, a leading First World nation with a dynamic market economy and flourishing democracy. Korean companies like Samsung, Hyundai, and LG are now household names across the world. In Lebanon, Haiti, and other hot spots around the globe, Korean peacekeepers serving under the UN flag strive to restore order and give shattered societies a chance to heal. Once dependent on international generosity, Korea is now an international aid donor, repaying its debt to the international community by giving nations less fortunate an opportunity to grow. ■

ABOVE: Scene of war damage in residential section of Seoul, Korea. The Capitol building can be seen in the background (right). (October 18, 1950, photo by Sgt. 1st Class Cecil Riley, Army)

Seoul's Han River today
(Photo by Korea Tourism Organization)

Han River in 1950
(Photo from Yonhap News)

Miracle on the Han River

While Korea's dramatic socioeconomic development between 1961 and 1996 was virtually unprecedented in the annals of human history, the "Miracle on the Han River," as that growth is called, was really no miracle. It was the product of sound government policy, great sacrifice on the part of the Korean worker, and dogged determination of the Korean people to raise their nation from poverty and ruin so that it could take its place among the world's leading nations.

The Miracle on the Han River began in 1961, when President Park Chung-hee took power. Possessing an iron will to develop Korea, Park set the nation on the path of export-led growth, beginning first with light industrial products and later moving to heavy industrial goods such as steel, cars, and ships. Bearing long work hours at initially low wages, Korean workers toiled to build a brighter future for their country. Park also pushed an ambitious initiative to modernize the Korean countryside. Called the New Community Movement, or Saemaeul Movement, the push led to dramatic improvements in rural living standards. By the end of Park's rule in 1979, Korea had been transformed from a society of agrarian peasants to an industrial one with a growing middle class.

Park's successors continued his market-friendly policies, pursuing progressively greater levels of liberalization. When Park took power, Korea's GDP per capita was just US$80; by 2007, it had reached US$20,000. By 2011, Korea's nominal GDP was in excess of US$1.1 trillion and the 15th highest in the world. As if to make the miracle official, Korea joined the OECD in 1996 and the G20 in 2010, bringing international recognition to Korea's transformation into a developed nation. ■

K O R
MORNING CALYPSO

OPPOSITE PAGE TOP: **The T-50 Golden Eagle, an indigenously developed, supersonic trainer manufactured by Korea Aerospace Industries. The Korean Air Force has operated the aircraft since 2005, and the Indonesian Air Force has ordered 16 of the trainers, with delivery set for 2013.**
OPPOSITE PAGE BOTTOM: **Hyundai automobiles await loading onto cargo ships at the port of Ulsan.** (Photos by Yonhap News)
TOP: **The Korean shipbuilding industry is the largest in the world.** (Photo by Yonhap News)
BOTTOM: **Worker oversees production at POSCO's massive integrated steel mill in Pohang, southeast Korea.** (Photo courtesy of POSCO)

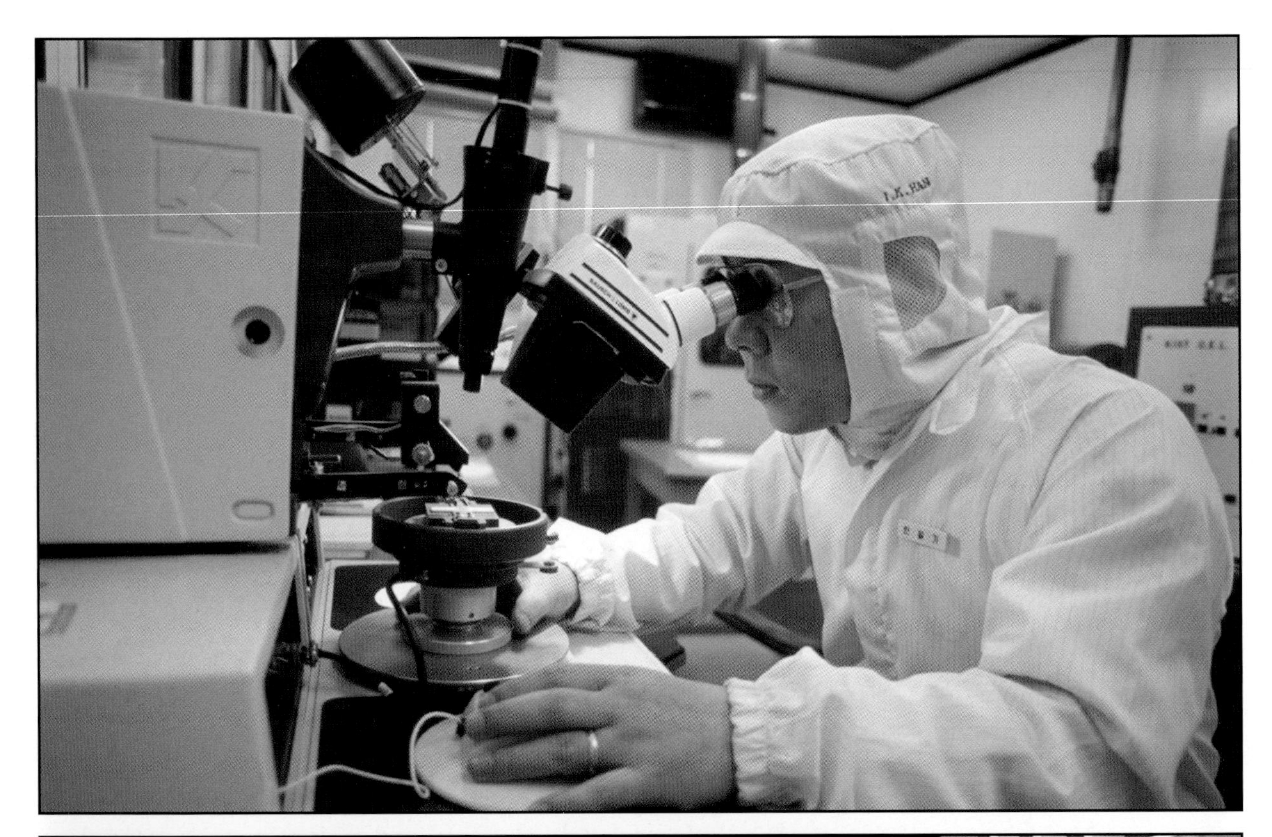

TOP: A semiconductor researcher at Korea Institute of Science of Technology (KIST), which has played a leading role in the development of Korean technology.
BOTTOM: Tourists examine LPG and electric hybrid cars made by Hyundai at Korea Energy Show, held at COEX in Seoul.
(Photos by Yonhap News)

On January 30, 2013, Korea successfully launched its Naro-1 rocket into space. (Photo courtesy of Korea Aerospace Research Institute)

Demographics

Population

Total Population: 48,955,203 (July 2013 est.)

Birth rate: 8.33 births/1,000 population (2013 est.)

Death rate: 6.5 deaths/1,000 population (2013 est.)

Urbanization

Urban population: 83% of total population (2010)

Rate of urbanization: 0.6% annual rate of change (2010-15 est.)

Major cities

Seoul (capital) 9.778 million; Busan 3.439 million; Incheon 2.572 million; Daegu 2.458 million; Daejeon 1.497 million (2009)

Religions

Christian 31.6% (Protestant 24%, Roman Catholic 7.6%), Buddhist 24.2%, other or unknown 0.9%, none 43.3% (2010 survey)

Literacy

Definition: age 15 and over can read and write

Total population: 97.9%

Male: 99.2%

Female: 96.6% (2002)

Economy

GDP (purchasing power parity in 2012 est.): $1.611 trillion.

Real GDP growth rate: (2010 est.) 6.3%; (2011 est.) 3.6%; (2012 est.) 2%

GDP per capita (current U.S. $): (2011 est.) $31,700; (2012 est.) $32,400

Unemployment rate (2012 est.): 3.2%

Inflation rate (consumer prices): (2011 est.) 4%; (2012 est.) 2.2%

Natural resources: coal, tungsten, graphite, molybdenum, lead, hydropower potential

Agriculture

Products: rice, root crops, barley, vegetables, fruit, cattle, pigs, chickens, milk, eggs, fish

Arable land: 16.58% of land area

Industry: electronics, telecommunications, automobile production, chemicals, shipbuilding, steel

Trade

Exports (2012 est.): $552.6 billion: semiconductors, wireless telecommunications equipment, motor vehicles, computers, steel, ships, petrochemicals

Imports (2012 est.): $514.2 billion: machinery, electronics and electronic equipment, oil, steel, transport equipment, organic chemicals, plastics

Major export markets (2011 est.): China (24.4%), U.S. (10.1%), Japan (7.1%)

Major importers to the Republic of Korea (2011 est.): China (16.5%), Japan (13%), US (8.5%), Saudi Arabia (7.1%), Australia (5%)

(Source: The CIA World Factbook , May 2013)

TOP: Forest of Korean flags. (Photo by Yonhap News)
BOTTOM: Dancers perform the Drum Dance, one of Korea's many beautiful traditional dances.
(Photo by Korea Tourism Organization)

Downtown Seoul at night. (Photo by Seoul Selection)

National Pride

With 5,000 years of history, Koreans are rightfully proud of their cultural heritage. Some of their cultural accomplishments include:

Hangeul - Writing System

- Invented in the mid-15th century under the rule of King Sejong the Great.
- UNESCO established the King Sejong Literacy Prize in 1988, an annual prize offered to an individual or group that contributes to the eradication of illiteracy worldwide.

Taegeukgi – Flag

- Symbolizes the basic ideas of East Asian cosmology.
- Perfect balance represented through the yin-yang symbol.
- Trigrams represent heaven, earth, water, and fire.

Aegukga – National Anthem

- National identity is represented through the territorial references to the East Sea, Mt. Baekdusan, and *mugunghwa* (the Rose of Sharon) national flower.

Taekwondo – National Sport

- Translates as "the way of the foot and fist."

Monk displays Tripitaka Koreana (scriptures) at Haeinsa Temple.
(Photo by Korea Tourism Organization)

14

Cargo being loaded at Busan, Korea's largest port. As of 2011, Korea was the ninth largest trading state, conducting US$1 trillion in annual trade. This has turned Korea into a global economic powerhouse alongside other developed nations like the United Kingdom and France. (Photo by Yonhap News)

CHAPTER 14

Envisioning a New Korea

Korea's transformation following the Korean War was all-encompassing. Global attention has traditionally focused on Korea's economic development, and not without reason. Korea's rapid economic growth beginning in the early 1960s and culminating with Korea's accession to the OECD in 1996 was almost without parallel; between 1962 and 1989, Korea's GDP grew at an incredible clip of 8% per annum. Wealth distribution was remarkably even during the growth years, leading to dramatic improvements in both urban and rural living standards. By the late 1990s, Korea had achieved per capita incomes and living standards on par with other developed nations in Europe, North America, and East Asia. The very nature of Korea's economy transformed too, moving away from subsistence agriculture in favor of production of consumer and industrial goods. In the early 20th century, Korea's major exports were rice and seaweed; by the end of the century, it was exporting cell phones, semiconductors, automobiles and oil tankers.

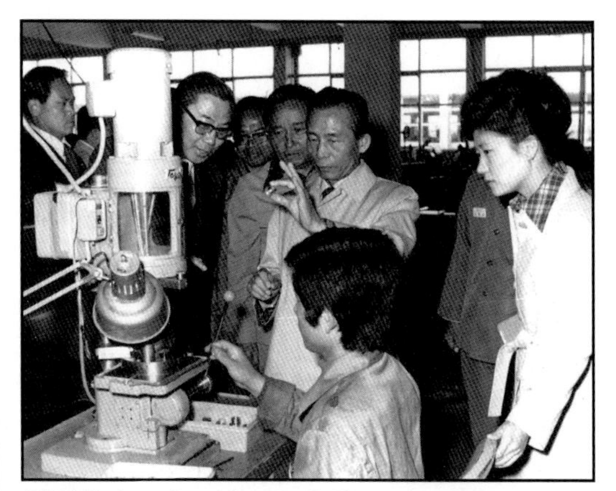

ABOVE: Late President Park Chung-hee inspects an electronics factory in Changwon Industrial Park in 1974. Standing next to him is his daughter Park Geun-hye, Korea's current president. (Photo by Yonhap News)

Occurring parallel to Korea's economic development was its political development. The two went hand-in-hand, in fact. As Korea transformed from a rural, agrarian society into a complex urban and industrialized one under authoritarian leadership, political attitudes grew increasingly sophisticated. Responding to the demands of the people for greater freedom, Korea's authoritarian rulers stepped aside in 1987 to hold multi-party elections. In 1998, Korea's democratic development achieved another milestone when Kim Dae-jung became the first opposition leader to be elected president. Subsequent reforms have strengthened the power and independence of Korea's legislative and judiciary branches and democratized local governments. In 2012, Korea actually placed ahead of the United States on The Economic Intelligence Unit's authoritative Democracy Index, ranking 20th worldwide. ■

U.S. Ambassador Samuel D. Berger (right) and Korean Minister of Transportation Kim Yun-Ki drive gold-colored memorial spikes on the new Hwangji Railroad line during dedication ceremonies. To Koreans, this new rail line meant more food, coal, and saved forests.
(Republic of Korea, 1964, photo by the NARA)

TOP: **Opening in July 1, 1970, the Seoul–Busan Expressway made it possible to get anywhere in Korea in just a single day and played a key role in Korea's industrialization.**
BOTTOM: **The New Village Movement in 1972.** (Photos by Yonhap News)

RIGHT: The Seoul–Busan Expressway. When it was completed in 1970, the expressway made it possible to travel between Seoul and Busan in just a single day, spurring Korea's economic development.
(Photo by Yonhap News)
OPPOSITE PAGE TOP: Incheon's Songdo International Business District is the centerpiece of the Incheon Free Economic Zone. One of the world's most technologically advanced cities, Songdo is also one of the world's most environmentally friendly urban areas and the headquarters of the UN Green Climate Fund.
(Photo by Seoul Selection)
OPPOSITE PAGE BOTTOM: Gwangan Bridge shines above the waters of Busan.
(Photo by Korea Tourism Organization)

TOP: Happy children run on the green lawn in front of Seoul City Hall.
BOTTOM LEFT: Foreign-born wives and their children. Korea is an increasingly multicultural society, and the Korean government has been enacting a variety of policies to promote diversity and social cohesion, including support for multicultural families.
BOTTOM RIGHT: A Korean university graduation ceremony. The much-lauded Korean passion for education is the basis upon which Korea's dramatic economic and social development was built. (Photos by Yonhap News)

Korea's religious environment is characterized by diversity and harmony.
TOP LEFT: Catholic mass at Myeong-dong Cathedral.
TOP RIGHT: Buddhist Lotus Lantern Festival in Seoul.
MIDDLE LEFT: 2012 meeting of various Korean religious leaders to promote interfaith harmony.
MIDDLE RIGHT: Joint Protestant worship session by National Council of Churches in Korea and Christian Council of Korea.
BOTTOM: The Confucian royal ancestral rite at Jongmyo Shrine, Seoul. (Photos by Yonhap News)

15

Logo of the 2010 G20 Seoul Summit, hosted in November 2010.
(Photo by Yonhap News)

CHAPTER 15

A Global Leader

Korea's social, economic, and political transfiguration has allowed it to take a place among the world's leading nations.

Korean corporations such as Samsung, Hyundai, and LG are global leaders in their fields. Korean companies have earned particular renown in high-tech industries such as semiconductors, next-generation displays, and information technology. In recent years, Samsung has achieved international success as the world's largest producer of smart phones. The transformation of automaker Hyundai from late-night-talk-show punchline to producer of some of the world's highest quality cars mirrors the transformation of Korea itself. Hyundai Heavy Industries, Samsung Heavy Industries, and Daewoo Shipbuilding and Marine Engineering have made Korea the world's largest shipbuilder. In 2009, Samsung C&T finished work on Dubai's 163-story Burj Khalifa, the world's tallest building.

Korea's successful hosting of major international events in recent years further demonstrates the nation's solidifying stature. In November 2010, the leaders of the world's largest economies converged on the Korean capital for the G-20 Seoul Summit. Likewise, Korea hosted the 2012 Nuclear Security Summit and the 2005 APEC Summit. Korea's role as an international venue has extended beyond the political, too. In 1988, Korea announced its place in the world by successfully hosting the Summer Olympic Games. The eyes of the sporting world were once again fixed on Korea in 2002, when the nation hosted the 2002 FIFA Korea-Japan World Cup, during which the Korean national soccer team pulled off a miraculous run to the semifinals. The Olympic spirit returns to Korea once again when the town of Pyeongchang hosts the Winter Olympic Games in 2018. ■

ABOVE: Heads of state at the 2012 Seoul Nuclear Security Summit. (Photo by Yonhap News)

TOP: Samsung Electronics' digital board in the heart of New York's Times Square attracts a diverse range of people from all around the world.
BOTTOM: On March 14, 2013, Samsung Electronics unveiled the Galaxy S4 at the Samsung Unpack event in Radio City Music Hall in New York City.
(Photos by Samsung)

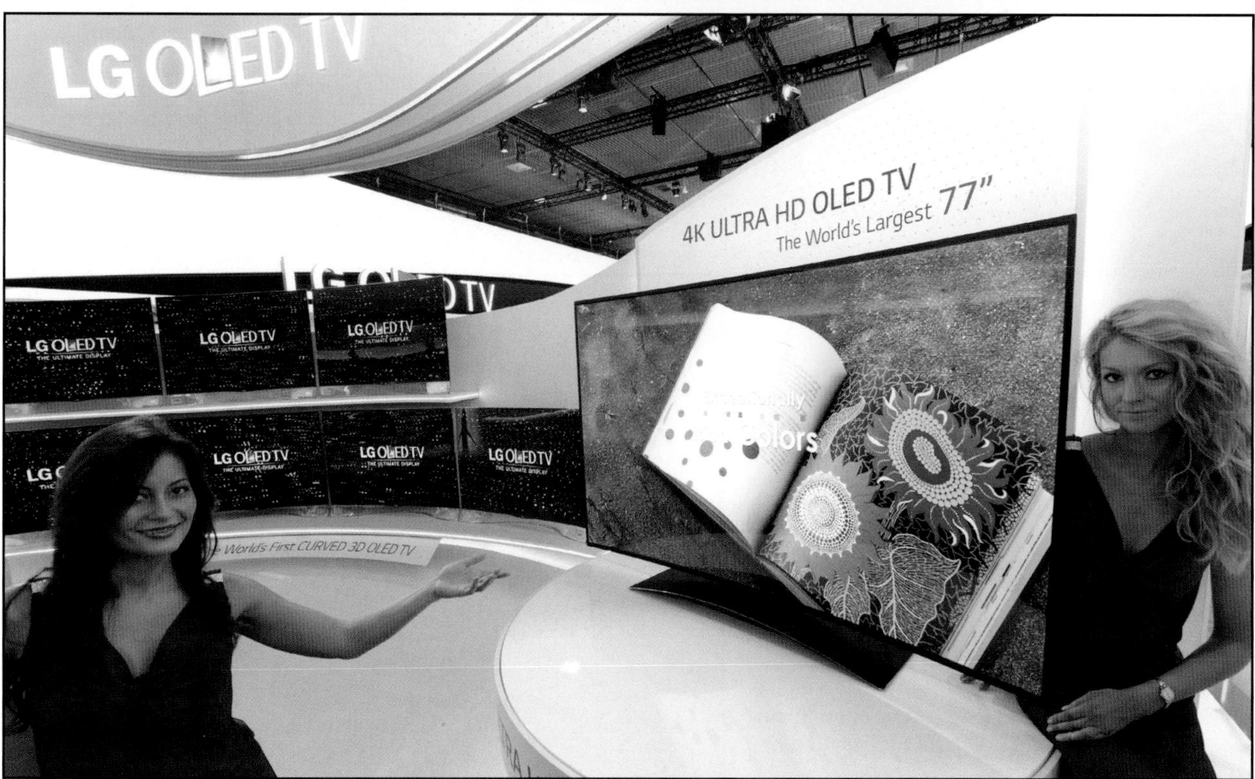

TOP: Hyundai Motor's HND-9 Venace at the Seoul Motor Show at Goyang's KINTEX exhibition center in March 2013. Some 384 companies from 14 nations participated in the show.
BOTTOM: On September 6, 2013, LG Electronics unveiled the world's largest Ultra HD curved screen OLED (organic light emitting diode) TV at the 2013 IFA international electronics show in Berlin. (Photo by LG)

ABOVE: An Airbus A380 from Korean Air takes flight. Korean Air is Korea's flagship airline.
RIGHT: Asiana is Korea's second major international air carrier. (Photos by Yonhap News)

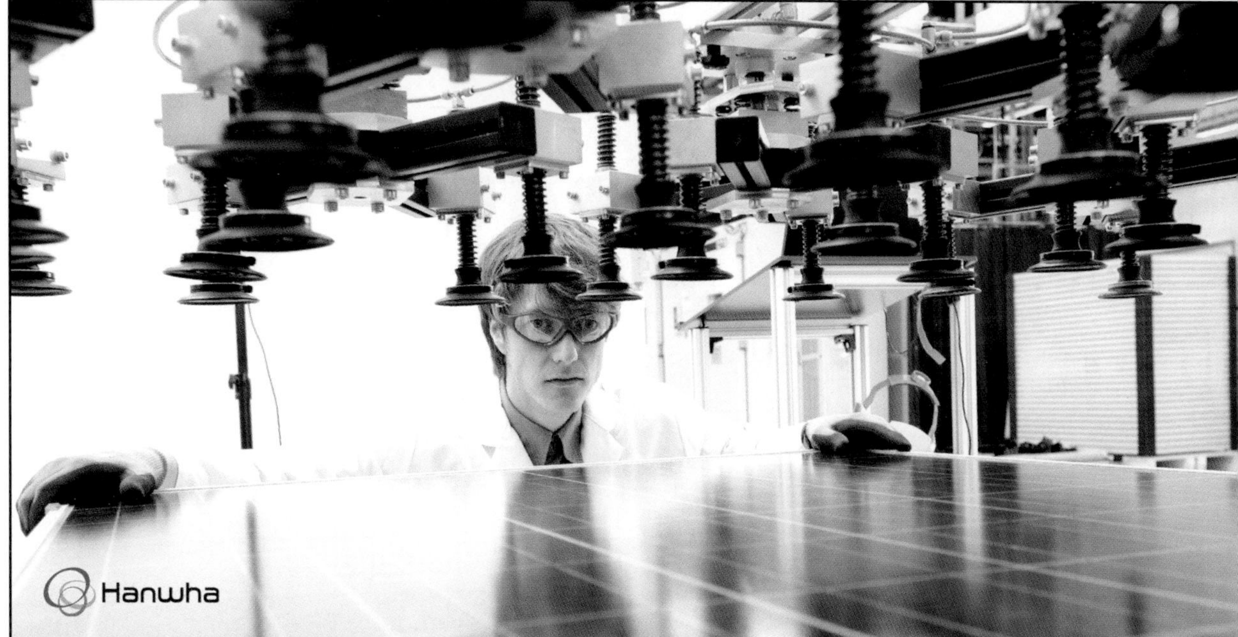

OPPOSITE PAGE TOP: SK oil refinery in Ulsan, southeast Korea. SK Energy is Korea's largest energy company, and its Ulsan refinery is one of the largest such facilities in the world. (Photo by SK)
OPPOSITE PAGE MIDDLE: GS Caltex, established in May 1967 as the first private oil company in the Republic of Korea, has opened a new chapter in oil industry history. With competitive power in petroleum, petrochemical, and lubricant production facilities, GS Caltex fulfilled the nation's need for petroleum energy and has been a strong driving force in Korea's economic development for more than 45 years. (Photo by GS)
TOP CENTER: LOTTE World Tower, Seoul (scheduled for completion in 2016). LOTTE has pursued continuous innovation and growth since its founding in 1967. It has built up the Group's capabilities by expanding its businesses in various fields including food, retail, tourism, petrochemicals, construction, and finance.
TOP RIGHT: LOTTE Department Store, Seoul. (Photos by LOTTE)
BOTTOM: German-based Hanwha Q CELLS is part of the South Korean Top Ten enterprise Hanwha Group and one of the world's leading photovoltaic (PV) companies. (Photo by Hanwha)

ABOVE: Korean-born World Bank President Jim Yong Kim with U.S. Secretary of State Hilary Clinton and President Barack Obama.
RIGHT: UN Secretary-General Ban Ki-moon. A career diplomat and formerly Korea's foreign minister, Ban became Secretary-General of the United Nations in 2007. As UN leader, he has strived tirelessly to bring peace to the world's trouble spots.

ABOVE: **President Park Geun-hye speaks during her inauguration, February 25, 2013. Korea's 18th president and first female president, Park is a rising global leader who was recently chosen as one of the world's 100 most influential people by American magazine TIME.** (Photo by Yonhap News)

ABOVE: Some 1,100 students form the Seoul Olympic emblem. The 1988 Summer Olympic Games, hosted in Seoul, introduced Korea and its development to the international community.

TOP: **Daegu Stadium, the venue of the 2011 World Championships in Athletics. Some 6,000 people from 212 nations, including athletes, staff, and journalists, participated in the event.** (Photo by Korea Tourism Organization)

LEFT: **Korean figure skater Kim Yu-na. One of Korea's best known sports stars, Kim won gold at the ladies single event in the 2010 Winter Olympic Games and holds several international figure skating records.** (Photos by Yonhap News)

ABOVE RIGHT: **At the International Olympic Committee (IOC) session in Durban, South Africa, in July 2011, IOC president Jacques Rogge announces the selection of the Korean town of Pyeongchang as the host of the 2018 Winter Olympic Games.**

Huge crowds turn out in front of Seoul City Hall to celebrate the 2002 FIFA World Cup, hosted in Korea and Japan.
(Photo by Korea Tourism Organization)

삼성화재
삼성화재
자동차보험
삼성화재
축구 사랑! 인권 사랑!
korea!
대한민국

Fans cheer on Korean boy band Super Junior at the SM Town Live concert at Le Zénith de Paris. (Photos by Yonhap News)

K-Pop and the Korean Wave

Korean pop culture has enjoyed tremendous popularity since 2000, initially in East Asian markets and expanding later to the Middle East, Europe, and the Americas. Dubbed the "Korean Wave," this cultural phenomenon reached its most recent apex in 2012 when rapper Psy's hit "Gangnam Style" became an overnight hit, earning the singer a performance before U.S. President Barack Obama.

At the forefront of the Korean Wave has been "K-pop," or Korean pop music. The development of information technology has allowed Korean pop groups to win enthusiastic followings overseas. A 2011 K-pop concert in Paris, for instance, sold out in just 10 minutes, prompting organizers to put on a second show. Girl and boy bands like Girls Generation, Wonder Girls, Big Bang and Super Junior have become household names in not just their native land, but also in Japan, China, and beyond.

There's more than just music to the Korean Wave, however. Korean TV dramas have grown especially popular, especially in Japan, China, and Southeast Asia. The 2002 drama "Winter Sonata," for instance, turned Korean actor Bae Yong-jun into a superstar in Japan and sparked a craze for all things Korean. Korean cinema, too, has earned a worldwide following, with Quentin Tarantino naming three Korean films to the list of his 20 favorite films released since 1992. The popularity of Korean film is further evidenced by the fact that three Hollywood productions in 2013 were directed by Korean directors.

The popularity of Korean pop culture has sparked interest in other aspects of Korean culture, too. Korean cuisine has achieved a worldwide presence, with Korean dishes like *bibimbap*, *bulgogi*, and, of course, *kimchi* earning fans across the globe. More and more people are visiting Korea, too. In 2012, over 11 million tourists came to Korea to take in its many sites, sounds, aromas, and tastes. In March 2013, inbound visitors to Korea surpassed 1 million, a single-month record. ■

The K-pop bands, which include (from far left) T-ara, Mblaq and T-Max, gathered on stage for the finale to sing Michael Jackson's Heal The World. ST PHOTOS: SAM CHIN

K-pop for a good cause

Girl group Afterschool (left) showed off their long lean legs in their short shiny dresses

new breed of Korean pop

Trend Spot
Sweet taches

Trends

K-POP CRAZY

Mass interest on social networking sites and a kitsch designer image make South Korean pop the latest fixation for London's teens, says Victoria Stewart

Generation performing in Tokyo this year Photograph: Toshifumi Kitamura/AFP

ers have a more Asian feel to them. That appeals to me. And their fans here appreciate their attempts to learn Japanese."

The linguistic nod makes commercial sense. Japan is the second biggest music market in the world after the US, with a 22% global share, according to the International Federation of the Phonographic Industry. CD sales in South Korea are one 30th of those in Japan.

K-pop artists have taken localisation to its natural conclusion: in their native South Korea, Toho Shinki are known as Dong Bang Shin Ki, while the rest of the world knows them as TVXQ. Girls' Generation perform in Japan as Shojo Jidai, and in South Korea as So Nyuh Shi Dae.

Stephen McClure, the Tokyo-based editor of McClure's Asia Music News,

Pop goes the whistle again

TIGER
it stands
FIRM

South Korea
Capital: Seoul
All about SOUTH KOREA

ABOVE: Headlines of Korean Pop Culture.

TOP: Korean rapper Psy is one of the most famous Koreans in the world thanks to his 2012 smash hit "Gangnam Style."
BOTTOM: K-pop fan club holding a flash mob event in the Palermo district of Buenos Aires, Argentina, to demand a performance by K-pop singers. (Photos by Yonhap News)

LEFT: Fashion by designer Lie Sang-bong. **One of the rising stars of the fashion world, Lie has made his name by harmonizing Parisian fashion with the beauty of the East. Particularly striking is his use of Korea's indigenous writing system, Hangeul, in his designs. The many celebrities he has dressed include fashionistas like Beyonce and Lady Gaga.**
(Photo by Yonhap News)

TOP RIGHT: **A bowl of rice mixed with seasoned vegetables and other ingredients, the *bibimbap* is one of many Korean dishes growing popular overseas.**
(Photo by Korea Tourism Organization)

ABOVE: **Laser Show at Gwangandaegyo Bridge, Busan.**
(Photo by Korea Tourism Organization)

16

Ceremony to mark the completion of a Korean-funded rural community development project outside of Dhaka, Bangladesh. Launched in 2010, the project was run by the Korea International Cooperation Agency (KOICA), which handles Korea's grant aid programs.
(Photo by Yonhap News)

CHAPTER 16

From Aid Recipient to Donor

Thirty-five years of Japanese colonial rule, national division, and the Korean War left Korea devastated. Fortunately, the international community—including the United States—was there to help. International aid played an important role in helping Korea get back on its feet after the war and paved the way for Korea's future prosperity. From 1945 to the 1990s, Korea received US $12 billion in foreign aid.

Through the early 1960s, aid grants—especially from the United States—were used to curb inflation, achieve financial stability, and invest in industrial facilities and other infrastructure. Later on, loans from international financial institutions such as the World Bank helped fund infrastructure development and other critical investments. On a more local level, volunteers from the U.S. Peace Corps brought valuable skills and enthusiasm, and private and church-based aid groups lent a helpful hand, too.

The Miracle on the Han River eventually rendered international aid unnecessary; in 1995, Korea was removed from the World Bank lending list. However, Koreans have never forgotten the generosity of the international community. In 2010, Korea joined the OECD's Development Assistance Committee, becoming the first former aid recipient to join the ranks of the world's major aid donors. In 2011, Korea was the world's 17th largest aid giver, providing a total of US$1.32 billion in official development assistance. Through the World Friends Korea program, Korea also sends thousands of volunteers abroad to bring needed skills and, just as importantly, Korea's unique development experience to the developing and emerging nations. Private Korean organizations and individuals engage in a variety of humanitarian and development activities overseas as well. ■

ABOVE: **U.S. aid enters the port of Busan in 1961.** (Photo by Yonhap News)

Korea's ODA Contributions

In 2011, Korea contributed US$1.32 billion in official development assistance (ODA), including US$970 million in bilateral aid and US$350 million in multilateral aid. Grants accounted for 57.5% of bilateral assistance, with loans accounting for the remaining 42.5%. Korea has pledged to triple its ODA contributions to US$3 billion or 0.25% of Gross National Income by 2015. ■

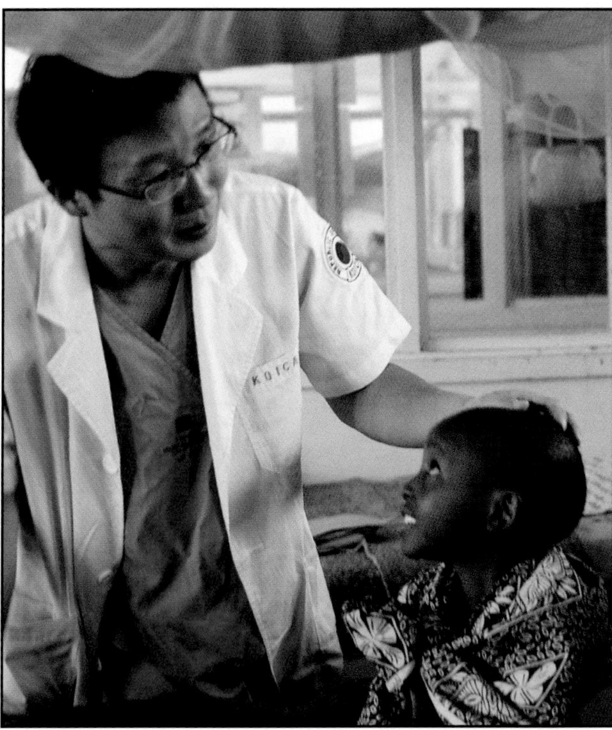

KOICA and EDCF

The Korea International Cooperation Agency (KOICA) was founded in 1991 to maximize the effectiveness of Korea's overseas assistance efforts. In particular, KOICA is charged with overseeing the distribution of Korea's grant aid. It maintains 30 offices in 28 partner countries in Asia, the Middle East, the former Soviet Union, Africa, and Latin America. Major aid themes include education; health; governance; agriculture, forestry, and fisheries; information and communication technologies; industry and energy; environment; disaster relief; achieving Millennium Development Goals; and climate change. KOICA achieves these goals through a variety of programs, such as aid projects, training programs, the dispatching of volunteers, and civil society cooperation.

In recent years, KOICA's efforts have expanded beyond poverty relief to focus increasingly on sustainable development, strengthening international partnerships, and ensuring local residents enjoy maximum benefit from aid.

Korea's loan aid, meanwhile, is handled by the Economic Development Cooperation Fund (EDCF), which in turn is executed through the Export-Import Bank of Korea. Korea's own experience as a major recipient of loan aid in the 1960s and 1970s have made Korea uniquely qualified as a loan aid provider. The EDCF provides development project loans, equipment loans, public-private partnership loans, two-step loans and community loans. In 2011, Vietnam alone received US$1.67 billion in loans for 43 projects, including a US$200 million loan to build the Lo Te–RachSoi highway. ■

Korea in Peacekeeping and International Security

Always mindful that the international community came to its defense in its hour of greatest need, Korea has been an active participant in UN peacekeeping operations. As of 2012, Korea had dispatched 639 men and women to participate in UN peacekeeping operations overseas, making it the 32nd most active participate in UN peacekeeping. Major recent deployments have included the dispatch of 359 personnel to southern Lebanon in 2007 and the sending of 242 personnel to Haiti in 2010. Korea also recently decided to send peacekeepers to South Sudan.

Korean troops have helped not only bring peace and stability to regions in need, but also served as conduits for humanitarian assistance and cultural exchange. For instance, Korea's Dongmyeong Unit—deployed to Lebanon since 2007—provides medical care to Lebanese within its area of operations and has opened popular classes for computer science, taekwondo, and Korean language. ■

BELOW RIGHT: Korean troops with the Zaytun Division hand out candy to local children in the northern Iraqi town of Arbil in 2004. The division was deployed to Iraq to carry out peace and reconstruction missions.
BOTTOM: Korean peacekeepers with the Dongmyeong Unit carry out recon in its area of operations in Lebanon.
(Photos by Yonhap News)

SECTION V

THE STRENGTH OF A NEW ALLIANCE

An honor guard performs at a ceremony to mark the anniversary of the establishment of the Combined Forces Command, a symbol of the Korea-U.S. alliance. The Combined Forces Command was formed on November 7, 1978, to handle the rapidly changing situation on the Korean Peninsula and boost the combined operational capacity of Korean and U.S. forces.
(Photo by Yonhap News)

CHAPTER 17

The Alliance Forged in Blood

The United States was the first Western nation to enter into diplomatic relations with Korea, signing a trade and friendship treaty with Korea in 1882. The friendship between the Korean and American peoples grew during the Korean War, when Korean and American soldiers fought side-by-side to protect the young Republic of Korea from Communist aggression. In October 1953, Korea and the United States signed a mutual defense pact, an alliance forged in the blood of all those who gave their lives in defense of freedom in the Korean War.

In the 60 years since it has come into existence, the Korea–U.S. alliance has become one of the closest military alliances in the world and remains critical to both Korean security and the protection of U.S. interests in East Asia. Korea today hosts 28,500 U.S. troops who play a key role in deterring North Korean aggression and maintaining peace and stability in Northeast Asia. Korean and U.S. forces work together under the Combined Forces Command, or CFC, a unique partnership ensuring efficient command and control in the event of an emergency on the Korean Peninsula.

Eager to repay its debt to the United States, Korea has consistently come to its ally's aid over the last six decades. Korea sent about 320,000 men to Vietnam to fight alongside the Americans in the Vietnam War; 5,000 Koreans were killed and 11,000 wounded. In recent years, Korea has sent substantial numbers of personnel to Iraq and Afghanistan to assist multinational efforts in those countries. Korean–American security cooperation is expanding to other theaters as well, including anti-proliferation, counter-terrorism, and piracy eradication. ■

ABOVE: USS *Blue Ridge* (LCC 19) Commanding Officer Capt. Thom W. Burke receives a flower wreath from Chon Yea Un, the daughter of a Republic of Korea naval officer, on behalf of the ROK Navy. (March 13, 2009, photo by MC3 Matthew D. Jordan)

"Closest Allies and Greatest Friends"

U.S. President Barack Obama has repeatedly praised Korea for its contributions to international peace and security. At the G-20 summit in London in 2009, President Obama called Korea "one of America's closest allies and greatest friends." Following another summit meeting with then-Korean President Lee Myung-bak on October 13, 2011, President Obama again hailed the unique alliance between Korea and the United States, saying:

> "This state visit reflects the fact that the Republic of Korea is one of our strongest allies. Because we've stood together, the people of South Korea, from the ruins of war, were able to build an economic miracle and become one of our largest trading partners, creating jobs and opportunity for both our peoples. Because we stood together, South Koreans were able to build a strong and thriving democracy and become a steady partner in preserving security and freedom not only on the Korean peninsula, but beyond.
>
> As I said this morning, this visit also recognizes South Korea's emergence as one of our key global partners. South Koreans have served bravely with us in Afghanistan and Iraq. South Korean forces have partnered with us to prevent piracy off the shores of Africa and stem the spread of weapons of mass destruction. Once a recipient of aid, South Korea has become a donor nation, supporting development from Asia to Africa. And under the President's personal leadership, Seoul served as host to the G20 summit last year and will host the next Nuclear Security Summit next year.
>
> South Korea's success is a tribute to the sacrifices and tenacity of the Korean people." ■

ABOVE: U.S. Army soldiers from Stryker Brigade Combat Team run out from an Infantry Carrier Vehicle in a live fire drill during joint exercises with Korea, dubbed Key Resolve and Foal Eagle, at Seungjin Fire Training Field in Pocheon, Korea, March 7, 2011.

ABOVE: The 97,000-ton USS *George Washington* of the U.S. Navy 7th Fleet enters the port of Busan to participate in joint exercises with Korea in July 2010. (Photos by Yonhap News)

TOP: **U.S. Army soldiers arrive at Daegu Air Base, South Korea, for Reception, Staging, Onward Movement, and Integration training exercise.** (DoD photo by Sgt. Albert Eaddy, U.S. Army)

ABOVE: **An M88A1 armored recovery vehicle hangs from a mobile crane as it is lifted onto a rail car in Busan.** (U.S. Army photo by Staff Sgt. Bryan Lewis)

RIGHT: **F-16 Fighting Falcons from the 35th and 80th Fighter Squadrons of the 8th Fighter Wing, Kunsan Air Base, Republic of Korea; the 421st Expeditionary Fighter Squadron of the 388th FW at Hill Air Force Base, Utah; the 55th EFS from the 20th FW at Shaw Air Force Base, S.C.; and from the 38th Fighter Group of the ROK Air Force, demonstrate an "Elephant Walk" as they taxi down a runway, during an exercise at Kunsan Air Base, Republic of Korea, March 2, 2012. The exercise showcased Kunsan Air Base air crews' capability to quickly and safely prepare an aircraft for a wartime mission.** (Photo by Senior Airman Brittany Auld, 8th Fighter Wing Public Affairs)

Monument dedicated to fallen U.S. and KATUSA soldiers

On June 8, 2012, U.S. and Republic of Korea military officials dedicated a monument to U.S. Military and the Korean Augmentation to the United States Army (KATUSA) troops who have died defending freedom in the Republic of Korea since the Armistice Agreement was signed.

The inscription on the monument, written in English and Korean, reads: *The people of the Republic of Korea have built this monument to honor the souls of the fallen soldiers of United States Forces Korea and KATUSAs who died fighting the communists here on the peninsula for the peace and democracy of the Republic of Korea. They fought until death to preserve the sacred spirit of liberal democracy that we are committed to pass on to our sons and daughters now and forever.*

"It has been almost 59 years since the armistice was signed, ending a devastating three-year war," said Combined Forces Command Commander General James D. Thurman at the dedication. "Even though the peninsula has been at relative peace since the signing, there have been a number of North Korean provocations; 43 Korean Augmentees to the U.S. Army and 92 U.S. service members where killed in the line of duty in Korea."

The KATUSA program, established during the early days of the Korean War, provides an opportunity for qualified Korean draftees who are fluent in English to apply for full-time duty in a U.S. Army unit. Those who apply for the program are chosen on random basis via lottery. Gen. Thurman noted that the program personifies the teamwork that keeps the U.S.-Republic of Korea alliance strong.

"Working together as a team has helped build mutual trust, common understanding, and cooperation between our countries, which is an inseparable bond we share today," he said. "Today we have more than 3,300 KATUSAs that continue to stand side-by-side with their U.S. partners as we deter aggression and preserve peace and stability on the peninsula." ■

TOP RIGHT: A monument dedication ceremony was held at the U.S. military base in Seoul to honor the U.S. Forces Korea and KATUSA who were killed in the line of duty while stationed in Korea following the 1953 Armistice. (Photo by Yonhap News)
BOTTOM RIGHT: Korean Army Academy cadets visiting the Combined Forces Command in Yongsan, Seoul, applaud USFK commander General Walter Sharp, February 3, 2010. (Photo by Yonhap News)

TOP: Veterans of the Korean War from several nations offer their respects at Seoul National Cemetery. (Photo by Yonhap News)
BOTTOM: The U.S. Marine Corps Silent Drill Platoon performs during the Battle of Changjin (Chosin) Reservoir Commemoration at the Korean National War Museum, November 10, 2010. (U.S. Marine Corps photo by Lance Cpl. Alejandro Peña)

Korean and American negotiators hold negotiations on the KORUS FTA in Seattle on September 7, 2006. (Photo by Yonhap News)

Economic Alliance

Sharing vibrant market economies, Korea and the United States enjoy a deep and flourishing economic relationship. Trade naturally accounts for much of this partnership. Korea is now the United States' seventh largest trading partner. Trade in goods between the two nations totaled US$101 billion in 2012; top U.S. exports to Korea included machinery, aircraft, and optical and medical instruments, while top Korean exports to the United States included cars, electrical machinery, and mineral fuel. There is large-scale trade in agricultural goods, services, entertainment, and other manufactured goods as well. Trade between Korea and the United States has only increased since the landmark Korea—U.S. Free Trade Agreement went into effect on March 15, 2012.

The "economic alliance" between Korea and the United States extends beyond just trade, however. Bilateral investment has created hundreds of thousands of jobs on both sides of the Pacific. American firms have long been among Korea's major foreign investors. U.S. direct investment in Korea totaled US$30.2 billion in 2010, led by the manufacturing and banking sectors. In recent years, however, Korean investment in the United States has increased dramatically. Korean direct investment in the United States totaled US$15.2 billion in 2010. Korean car manufacturer Hyundai's massive plant in Montgomery, Alabama, employs over 3,000 workers, while subsidiary car maker Kia rejuvenated the town of West Point, Georgia, when it opened a car plant there in 2009, creating 7,500 jobs in the area. ■

KORUS FTA

Korea and the United States signed the Korea–U.S. Free Trade Agreement (KORUS FTA) on June 30, 2007, with the agreement going into effect on March 15, 2012, following its ratification by the legislative bodies of both nations.

According to the Office of the United States Trade Representative, the KORUS FTA represents "the United States' most commercially significant free trade agreement in almost two decades." When it went into effect, 80% of bilateral exports of consumer and industrial goods became duty free; by 2016, the percentage will grow to 95%. By 2022, most of the remaining tariffs and quotas will be removed.

The agreement produced an almost instantaneous win-win effect. In 2012, exports of American manufactured goods increased by 1.3%. Korean exports to the United States increased by 1.4%. ■

Busan's Gamman Pier is busy even in the early morning hours. (Photo by Yonhap News)

TOP: President Park Geun-hye lays a wreath at the Tomb of the Unknown Soldier in Arlington National Cemetery. Park said, "As the President of the Republic of Korea, I once again thank those who sacrificed themselves in the Korean War. Without them, Korea's prosperity would have been impossible." Behind Park stands Maj. Gen. Michael S. Linnington, commander of the United States Army Military District of Washington. (Photo by Yonhap News)
BOTTOM: President Park visits the Korean War Veterans Memorial in Washington, D.C., with U.S. Secretary of Veterans Affairs Eric Shinseki (left) and Maj. Gen. Lee Seo-young (right), defense attaché of the Korean embassy in Washington, D.C.
(Photo by Yonhap News)

TOP: President Park receives a standing ovation after her address to a joint session of the U.S. Congress, May 7, 2013. (Photo by Yonhap News)
BOTTOM: President Park Geun-hye shakes hands with U.S. President Barack Obama after a joint press conference at the White House, May 7, 2013. (Photo by Yonhap News)

PARTNERS FOR PEACE AND PROSPERITY

Established in the aftermath of the Korean War, the Republic of Korea–U.S. alliance has served as the bedrock of Korea's security and regional stability for 60 years, and will continue to play a vital part in promoting global peace and prosperity in the 21st century.

Victory of the Alliance

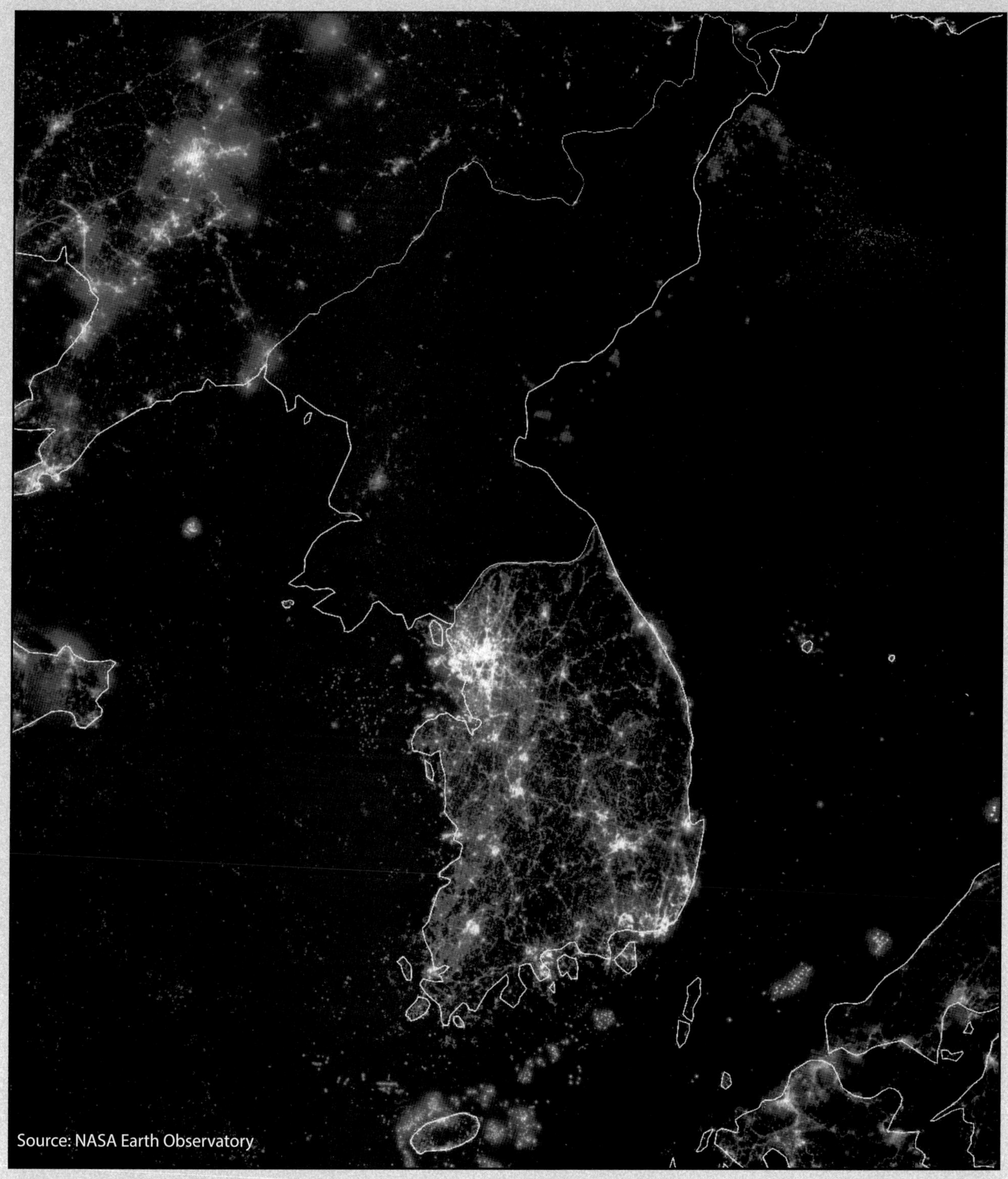

The economic contrast between the two Koreas is even more dramatic, as is best illustrated by a famous photograph of the Korean Peninsula at night, taken from space. Below the DMZ is a nation awash in lights, while above the DMZ there is almost nothing but darkness.

Taegeukgi at Exhibition Hall of the Gangneung Unification Park.
(Photo by Korea Tourism Organization)

CHAPTER 18

Defining Foreign Relations

Relations between South Korea and the United States, on one hand, and North Korea, on the other, have remain strained. The strong Korea–U.S. alliance has managed to deter major North Korean aggression since the conclusion of the Korean War, but North Korean provocations against both South Korea and the United States have continued. In 2010, North Korea launched a series of provocations against the South, most notably the sinking of the South Korean warship Cheonan and the shelling of the South Korean island Yeonpyeongdo. These acts of aggression—along with North Korea's internationally condemned nuclear weapons and long-range missile programs—serve to remind the peoples of both Korea and the United States for the need for close bilateral cooperation.

Despite repeated North Korean belligerence, both South Korea and the United States have continuously strived to modify North Korean behavior and improve relations with Pyongyang. Multilateral negotiations with North Korea through the so-called six-party talks over its nuclear and missile programs have been ongoing, albeit through fits and starts. U.S. humanitarian assistance to North Korea, largely in the form of food, energy, and medical aid, totaled US$1.3 billion between 1995 and 2008.

South Korean efforts towards reconciliation with the North have been even more ambitious. South Korean attempts at engagement—launched in 1998 under the banner of the "Sunshine Policy"—led to a marked cooling of tensions on the Korean Peninsula for roughly a decade, producing inter-Korean summit meetings in 2000 and 2007. Late South Korean President Kim Dae-jung's efforts to improve relations with North Korea earned him a Nobel Peace Prize in 2000. Continued North Korean aggression made the Sunshine Policy untenable and inter-Korean ties once again deteriorated, but the present government of President Park Geun-hye has restarted efforts to build trust between the two Koreas in an effort to bring lasting peace and prosperity to the Korean Peninsula. ■

ABOVE: South and North Korean troops face off at the truce village of Panmunjeom.
(Photo courtesy of MPVA)

ABOVE: North Korean soldiers in steel helmets and South Korean and U.S. soldiers stare at one another across the Military Demarcation Line. (Photo by Yonhap News)

North Korean soldiers use binoculars to keep an eye on the South, in Panmunjeom.
(Photo by Edward N. Johnson)

The DMZ

The Korean Demilitarized Zone, or DMZ, stretches 250 km across the Korean Peninsula, dividing the peninsula in two, roughly along the 38th Parallel. It serves as the border between North and South Korea, and despite its name, it is in fact one of the most militarized frontiers on the planet. U.S. President Bill Clinton once described it as one of the scariest places on Earth.

The DMZ follows the front line of the Korean War at the moment the Armistice Agreement was signed on July 27, 1953. Both sides pulled back two kilometers from the final line of contact, producing a four-km-wide no-man's land to serve as a buffer between the two warring sides. The Armistice Agreement bars the bringing of heavy weapons into the DMZ, but both sides are permitted to conduct regular patrols in their respective halves. Both North Korea and South Korea maintain a system of metal fences along the DMZ, along with a series of guard posts and observation towers. In the western sector of the DMZ is the truce village of Panmunjeom and the Joint Security Area (JSA), where the two sides sometimes come together to confer on matters pertaining to the Armistice. The iconic JSA—the world's last remaining Cold War frontier—is also a popular tourist destination.

Its name is not the only thing ironic about the DMZ. Despite its militarized reality, the DMZ may be the most tranquil location in Korea. Its status as a no-man's land has turned the inter-Korean frontier into a great nature reserve that is now home to many endangered species of plants and animals, including migratory cranes and leopards. ■

TOP: Rusting signs mark the Military Demarcation Line and the Bridge of No Return.
(Photo by Korea Tourism Organization)
ABOVE LEFT: Display area at the DMZ Museum.
(Photo by Korea Tourism Organization)
RIGHT: Milestone of the 38th Parallel line.
(Photo by Korea Tourism Organization)

South Korean soldiers patrol the iron fence of the Southern Limit Line, the southern boundary of the DMZ.
(Photo by Yonhap News)

THE REPUBLIC OF KOREA THANKS YOU

Korean preschool children meet Korean War veterans visiting Korea at the invitation of the MPVA.
(Photo courtesy of the MPVA)

The Republic of Korea Thanks You

The Republic of Korea recognizes the vital role that the United States and its outstanding veterans have played in its progress and success since the Korean War ended. The country has erected monuments, sent out Korean War Service medals, supported Veterans' associations, and established the Revisit Korea program, in which thousands of men and women who fought in the Korea War have been able to return to South Korea.

"We've helped a lot of other countries and done a whole lot of things, and boy, they've never really expressed their appreciation like the Koreans have," observes veteran Al Ortiz. "They just can't stop saying 'thank you.' Every time you turn around, they're doing something for the veterans. They have receptions, they give us medallions, they do all kinds of things, they're always doing. I think it's very nice, I really appreciate that."

The opportunity to revisit the country where they served has been a poignant, rewarding opportunity for the veterans who have returned to the country, which most remember as being devastated.

"It's hard to describe a nation that was completely ravaged, and then all of a sudden—skyscrapers," says veteran Bob Ramos. "I just don't have the words to describe the feeling that I—in a small portion—had something to do with it. Just my small part to help a country to be free."

Veteran Jay Staker echoes those comments. "They've made choices, too, and I think they've made a great choice. I'm proud of what they done. Vicariously, I've contributed very little to what they've done maybe. But boy, it gave them the opportunity to do what they wanted to do in their world and in their country. I'm proud to say, 'Look what they've done.' But also, secretly, I'm proud of what I've done too."

"It was a miracle," says veteran William Maloney. "The mythological phoenix rising from the ashes. It just was hard to believe. Seoul had one bombed-out bridge in the 1950s, and now I can't count them. The population of greater Seoul is probably about... 28 million plus. And they have a fast train that goes from Seoul to Busan, which is what we used to call Pusan. From a retrospective view, everyone talked about the forgotten war and the stalemate, but it wasn't a stalemate, it was a victory! The Korean War was the first battle of the Cold War, and it was bloody, but... we won! I take a few minutes every once in a while in talking with our veterans and I tell them that. That they won." ■

TOP: President Park Geun-hye exchanges greetings with American Korean War veterans during a visit to the Korean War Veterans Memorial in Washington, D.C., on May 6, 2013. (Photo courtesy of Cheong Wa Dae)
BOTTOM: The Korean group "Little Angels" performs in honor of U.S. Veterans at a special event honoring Korean War veterans in Washington, D.C., hosted by the Ministry of Patriots and Veterans Affairs. (Photo courtesy of the MPVA)

TOP: Visiting U.S. veterans lay a wreath and burn incense at the Memorial Monument (Hyeon Chung Tap) in the National Cemetery at Seoul. The Monument, a national shrine to Korea's Fallen, holds tablets that name 104,000 ROK soldiers whose remains were never recovered. It also contains the ashes of 6,000 unknown ROK soldiers.
(Photo courtesy of the MPVA)
BOTTOM: Korean troops take care of graves at Busan UN Cemetery.
(Photo by Yonhap News)

U.S. and the Republic of Korea reach mutual security pact
August 1953

Pres. Rhee resigns
April 27, 1960

Revisit program established
1975

Hyundai creates first indigenous car
1976

Combined Forces Command established
1978

Park Chung-hee assassinated
October 26, 1979

Chon Doo Hwan elected President
August 1980

New Constitution goes into effect
October 1980

Samsung introduces Korea's first personal computer
1982

South Korea establishes the Economic Development Cooperation Fund
1987

Voters approve new Constitution
October 1987

Roh Tae-Woo elected President
December 1987

Seoul hosts Summer Olympics
1988

South Korea adopts new Constitution
February 1988

U.N. admits South Korea
September 1991

Diplomatic relations established between China and South Korea
August 1992

Kim Il-Sung dies
July 1994

Agreed Framework Pact ends
1999

First-ever inter-Korea Summit
June 2000

South Korea cohosts World Cup
2002

Second inter-Korea summit
October 2007

South Korea admitted to G-20
2009

Kim Jong Il dies
December 2011

Samsung becomes world's leading cell-phone manufacturer
2012

US-Korea Free Trade Agreement goes into force
March 2012

Park Geun-hye elected President
February 2013

ABOVE: Special event honoring a group of U.S. veterans visiting the Republic of Korea as part of the Ministry of Patriots and Veterans Affairs Revisit Program. (Photo courtesy of KVA)

Revisiting Korea - Making the "Forgotten War" Unforgotten

In 1975, the Republic of Korea government established the Korean War Veterans Revisit Program. The government began the program to honor and show appreciation to all U.N. forces who fought to liberate the country from aggression during the Korean War. The Republic of Korea, through its Ministry of Patriots and Veterans Affairs, pays all expenses and half the transportation cost for veterans to visit Korea for five days.

During their visit, participants tour war memorials and battle sites around Incheon and Seoul, the Demilitarized Zone, an old Korean village, and the site where the armistice was signed in 1953. At the end of their visit, the MPVA holds a banquet in their honor and presents each veteran with an Ambassador for Peace Medal and several other gifts.

Since 1975, over 26,000 veterans have returned to the Republic of Korea to experience for themselves the transformation they helped create in the country they served. With the 60th anniversary, the government has expanded the program to include grandchildren of veterans, including a Peace Camps for Youth program, which furthers education about Korea and the profound impact of the Korean War veterans in South Korea.

"I think it's probably unique, the way South Korea has attempted to thank and help people in the revisit program," says veteran William McCulloch. "I've been to France and Germany and Italy and so on and they're all appreciative of U.S. help, but the way South Korea has set up this revisit program is unique. It's called a forgotten war, but I think the revisit program has done a great deal to keep the United States informed on what went on then and what is going on now. The South Koreans ought to be complimented for rebuilding their country and for their program of revisit. It does a lot to make the so-called 'Forgotten War' unforgotten."

TOP: Honorable Sungchoon Park, Minister of Patriots and Veterans Affairs, presents a 60th Anniversary Commemorative Medal to Congressman Charles B. Rangel, a Korean War veteran. (Photo courtesy of the MPVA)

MIDDLE: Chairman of the Joint Chiefs of Staff Adm. Mike Mullen greets soldiers after a tour of the DMZ in Panmunjeom. (U.S. Navy photo by Mass Communication Specialist 1st Class Chad J. McNeeley)

BOTTOM: Korean War monument at Incheon, commemorating one of the most significant and successful operations in the history of amphibious warfare. (U.S. Navy photo by Todd Macdonald)

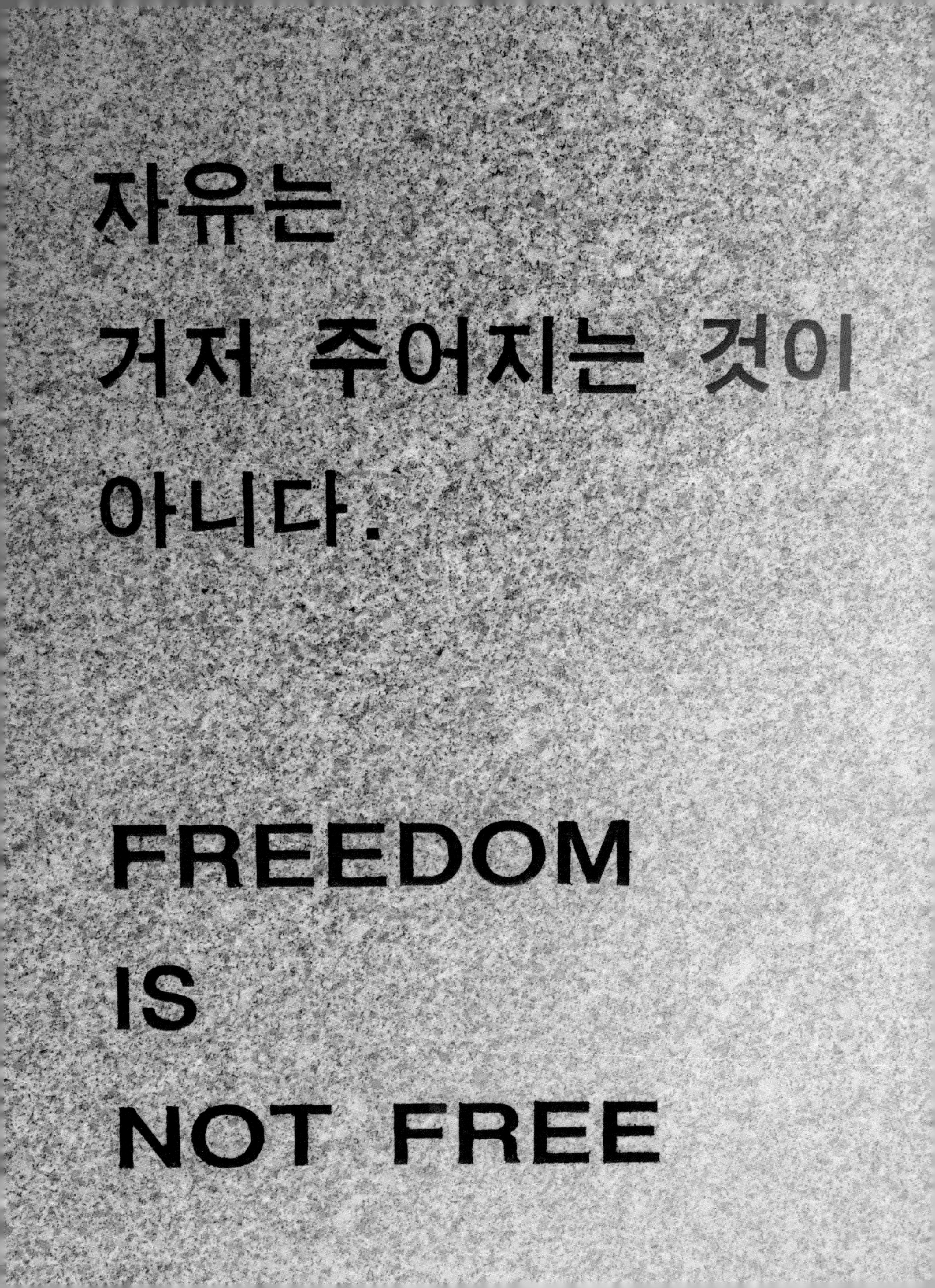

자유는

거저 주어지는 것이

아니다.

FREEDOM

IS

NOT FREE

KOREA REBORN

A GRATEFUL NATION

HONORS WAR VETERANS FOR 60 YEARS OF GROWTH